MILITARY AND VETERAN ISSUES

NAVY SUBMARINES

OVERVIEW OF VIRGINIA AND OHIO CLASS PROGRAMS

MILITARY AND VETERAN ISSUES

Additional books in this series can be found on Nova's website under the Series tab.

Additional e-books in this series can be found on Nova's website under the e-book tab.

MILITARY AND VETERAN ISSUES

NAVY SUBMARINES

OVERVIEW OF VIRGINIA AND OHIO CLASS PROGRAMS

TERRY COLEMAN
EDITOR

New York

Library of Congress Cataloging-in-Publication Data

ISBN: 978-1-63483-236-6

Published by Nova Science Publishers, Inc. † New York

CONTENTS

PREFACE

The U.S. Navy operates three types of submarines- nuclear-powered ballistic missile submarines (SSBNs), nuclear-powered cruise missile and special operations forces (SOF) submarines (SSGNs), and nuclear-powered attack submarines (SSNs). This book provides background information and issues for Congress on the Virginia-class SSN program. The Navy's proposed FY2016 budget requests $5,340.1 million in procurement, advance procurement (AP), and Economic Order Quantity (EOQ) funding for the program. Decisions that Congress makes on procurement of Virginia-class boats could substantially affect U.S. Navy capabilities and funding requirements, and the U.S. shipbuilding industrial base. Furthermore, this book provides background information and potential oversight issues for Congress on the Ohio replacement program (ORP), a program to design and build a new class of 12 ballistic missile submarines (SSBNs) to replace the Navy's current force of 14 Ohio-class SSBNs. The Ohio replacement program is also known as the SSBN(X) program. The Navy wants to procure the first Ohio replacement boat in FY2021, with advance procurement (AP) funding starting in FY2017. The Navy has identified the Ohio replacement program as its top priority program.

In: Navy Submarines
Editor: Terry Coleman

ISBN: 978-1-63483-236-6

Chapter 1

NAVY VIRGINIA (SSN-774) CLASS ATTACK SUBMARINE PROCUREMENT: BACKGROUND AND ISSUES FOR CONGRESS*

Ronald O'Rourke

SUMMARY

The Navy has been procuring Virginia (SSN-774) class nuclear-powered attack submarines since FY1998. The two Virginia-class boats requested for procurement in FY2016 are to be the 23rd and 24th boats in the class. The 10 Virginia-class boats programmed for procurement in FY2014- FY2018 (two per year for five years) are being procured under a multiyear-procurement (MYP) contract.

The Navy estimates the combined procurement cost of the two Virginia-class boats requested for procurement in FY2016 at $5,376.9 million or an average of $2,688.4 million each. The boats have received a total of $1,613.5 million in prior-year advance procurement (AP) funding and $416.9 million in prior-year Economic Order Quantity (EOQ) funding. The Navy's proposed FY2016 budget requests the remaining $3,346.4 million needed to complete the boats' estimated combined procurement cost. The Navy's proposed FY2016 budget also requests $1,663.8 million in AP funding and $330.0 million in EOQ funding for

* This is an edited, reformatted and augmented version of a Congressional Research Service publication RL32418, prepared for Members and Committees of Congress, dated March 26, 2015.

Virginia-class boats to be procured in future fiscal years, bringing the total FY2016 funding request for the program (excluding outfitting and post-delivery costs) to $5,340.1 million.

The Navy's proposed FY2016 budget also requests $167.7 million in research and development funding for the Virginia Payload Module (VPM). The funding is contained in Program Element (PE) 0604580N, entitled Virginia Payload Module (VPM), which is line 123 in the Navy's FY2016 research and development account.

The Navy plans to build Virginia-class boats procured in FY2019 and subsequent years with an additional mid-body section, called the Virginia Payload Module (VPM), that contains four large-diameter, vertical launch tubes that the boats would use to store and fire additional Tomahawk cruise missiles or other payloads, such as large-diameter unmanned underwater vehicles (UUVs). The Navy estimates that building Virginia-class boats with the VPM might increase their unit procurement costs by about 13%. It would increase the total number of torpedo-sized weapons (such as Tomahawks) that they could carry by about 76%. The Navy's FY2016 shipbuilding plan calls for building one of the two Virginia-class boats to be procured in FY2019, and one of the two Virginia-class boats to be procured in FY2020, with the VPM.

The Navy's FY2016 30-year SSN procurement plan, if implemented, would not be sufficient to maintain a force of 48 SSNs consistently over the long run. The Navy projects under that plan the SSN force would fall below 48 boats starting in FY2025, reach a minimum of 41 boats in FY2029, and remain below 48 boats through FY2036.

Potential issues for Congress regarding the Virginia-class program include the Virginia-class procurement rate in coming years, particularly in the context of the SSN shortfall projected for FY2025-FY2034 and the larger debate over future U.S. defense strategy and defense spending.

INTRODUCTION

This report provides background information and issues for Congress on the Virginia-class nuclear-powered attack submarine (SSN) program. The Navy's proposed FY2016 budget requests $5,340.1 million in procurement, advance procurement (AP), and Economic Order Quantity (EOQ) funding for the program. Decisions that Congress makes on procurement of Virginia-class boats could substantially affect U.S. Navy capabilities and funding requirements, and the U.S. shipbuilding industrial base.

The Navy's Ohio Replacement (SSBN[X]) ballistic missile submarine program is discussed in another CRS report.[1]

BACKGROUND

U.S. Navy Submarines[2]

The U.S. Navy operates three types of submarines—nuclear-powered ballistic missile submarines (SSBNs),[3] nuclear-powered cruise missile and special operations forces (SOF) submarines (SSGNs),[4] and nuclear-powered attack submarines (SSNs). The SSNs are general-purpose submarines that can (when appropriately equipped and armed) perform a variety of peacetime and wartime missions, including the following:

- covert intelligence, surveillance, and reconnaissance (ISR), much of it done for national-level (as opposed to purely Navy) purposes;
- covert insertion and recovery of SOF (on a smaller scale than possible with the SSGNs);
- covert strikes against land targets with the Tomahawk cruise missiles (again on a smaller scale than possible with the SSGNs);
- covert offensive and defensive mine warfare;
- anti-submarine warfare (ASW); and
- anti-surface ship warfare.

During the Cold War, ASW against the Soviet submarine force was the primary stated mission of U.S. SSNs, although covert ISR and covert SOF insertion/recovery operations were reportedly important on a day-to-day basis as well.[5] In the post-Cold War era, although anti-submarine warfare remains a mission, the SSN force has focused more on performing the other missions noted on the list above.

Attack Submarine Force Levels

Force-Level Goal

The Navy wants to achieve and maintain a fleet in coming years of 306 ships, including 48 SSNs.[6] For a review of SSN force level goals since the Reagan Administration, see *Appendix A*.

Force Level at End of FY2014

The SSN force included more than 90 boats during most of the 1980s, when plans called for achieving a 600-ship Navy including 100 SSNs. The number of SSNs peaked at 98 boats at the end of FY1987 and has declined since then in a manner that has roughly paralleled the decline in the total size of the Navy over the same time period. The 55 SSNs in service at the end of FY2014 included the following:

- 41 Los Angeles (SSN-688) class boats;
- 3 Seawolf (SSN-21) class boats; and
- 11 Virginia (SSN-774) class boats.

Los Angeles- and Seawolf-Class Boats

A total of 62 Los Angeles-class submarines, commonly called 688s, were procured between FY1970 and FY1990 and entered service between 1976 and 1996. They are equipped with four 21-inch diameter torpedo tubes and can carry a total of 26 torpedoes or Tomahawk cruise missiles in their torpedo tubes and internal magazines. The final 31 boats in the class (SSN-719 and higher) are equipped with an additional 12 vertical launch system (VLS) tubes in their bows for carrying and launching another 12 Tomahawk cruise missiles. The final 23 boats in the class (SSN-751 and higher) incorporate further improvements and are referred to as Improved Los Angeles class boats or 688Is. As of the end of FY2014, 21 of the 62 boats in the class had been retired.

The Seawolf class was originally intended to include about 30 boats, but Seawolf-class procurement was stopped after three boats as a result of the end of the Cold War and associated changes in military requirements. The three Seawolf-class submarines are the *Seawolf* (SSN-21), the *Connecticut* (SSN-22), and the *Jimmy Carter* (SSN-23). SSN-21 and SSN-22 were procured in FY1989 and FY1991 and entered service in 1997 and 1998, respectively. SSN-23 was originally procured in FY1992. Its procurement was suspended in 1992 and then reinstated in FY1996. It entered service in 2005. Seawolf-class submarines are larger than Los Angeles-class boats or previous U.S. Navy SSNs.[7] They are equipped with eight 30-inch-diameter torpedo tubes and can carry a total of 50 torpedoes or cruise missiles. SSN-23 was built to a lengthened configuration compared to the other two ships in the class.[8]

Virginia (SSN-774) Class Program

General

The Virginia-class attack submarine (see *Figure 1*) was designed to be less expensive and better optimized for post-Cold War submarine missions than the Seawolf-class design. The Virginia-class design is slightly larger than the Los Angeles-class design,[9] but incorporates newer technologies. Virginia-class boats currently cost about $2.8 billion each to procure. The first Virginia-class boat entered service in October 2004.

Past and Projected Annual Procurement Quantities

Table 1 shows annual numbers of Virginia-class boats procured from FY1998 (the lead boat) through FY2014, and numbers scheduled for procurement under the FY2016-FY2020 Future Years Defense Plan (FYDP).

Table 1. Annual Numbers of Virginia-Class Boats Procured or Projected for Procurement

FY98	FY99	FY00	FY01	FY02	FY03	FY04	FY05	FY06	FY07	FY08	FY09
1	1	0	1	1	1	1	1	1	1	1	1
FY10	FY11	FY12	FY13	FY14	FY15	FY16	FY17	FY18	FY19	FY20	
1	2	2	2	2	2	2	2	2	2	2	

Source: Table prepared by CRS based on U.S. Navy data.

Multiyear Procurement (MYP)

The 10 Virginia-class boats shown in *Table 1* for the period FY2014-FY2018 (referred to as the Block IV boats) are being procured under a multiyear procurement (MYP) contract[10] that was approved by Congress as part of its action on the FY2013 budget, and awarded by the Navy on April 28, 2014. The eight Virginia-class boats procured in FY2009-FY2013 (the Block III boats) were procured under a previous MYP contract, and the five Virginia-class boats procured in FY2004-FY2008 (the Block II boats) were procured under a still-earlier MYP contract. The four boats procured in FY1998-FY2002 (the Block I boats) were procured under a block buy contract, which is an arrangement somewhat similar to an MYP contract.[11] The boat procured in FY2003 fell between the FY1998-FY2002 block buy contract and the FY2004-FY2008 MYP arrangement, and was contracted for separately.

Source: U.S. Navy file photo accessed by CRS on January 11, 2011, at http://www.navy.mil/search/display.asp? story_id=55715.

Figure 1. Virginia-Class Attack Submarine.

Joint Production Arrangement

Virginia-class boats are built jointly by General Dynamics' Electric Boat Division (GD/EB) of Groton, CT, and Quonset Point, RI, and Newport News Shipbuilding (NNS), of Newport News, VA, which forms part of Huntington Ingalls Industries (HII).[12] Under the arrangement, GD/EB builds certain parts of each boat, NNS builds certain other parts of each boat, and the yards take turns building the reactor compartments and performing final assembly of the boats. GD/EB is building the reactor compartments and performing final assembly on boats 1, 3, and so on, while NNS is doing so on boats 2, 4, and so on. The arrangement results in a roughly 50-50 division of Virginia-class profits between the two yards and preserves both yards' ability to build submarine reactor compartments (a key capability for a submarine-construction yard) and perform submarine final-assembly work.[13]

Cost-Reduction Effort

The Navy states that it achieved a goal of reducing the procurement cost of Virginia-class submarines so that two boats could be procured in FY2012 for combined cost of $4.0 billion in constant FY2005 dollars—a goal referred to as "2 for 4 in 12." Achieving this goal involved removing about $400 million (in constant FY2005 dollars) from the cost of each submarine. (The Navy calculates that the unit target cost of $2.0 billion in constant FY2005 dollars for each submarine translates into about $2.6 billion for a boat procured in FY2012.)[14]

Virginia Payload Module (VPM)

The Navy plans to build Virginia-class boats procured in FY2019 and subsequent years (i.e., the anticipated Block V and beyond boats) with an additional mid-body section, called the Virginia Payload Module (VPM). The VPM, reportedly about 70 feet in length[15] (earlier design concepts for the VPM were reportedly about 94 feet in length),[16] contains four large-diameter, vertical launch tubes that would be used to store and fire additional Tomahawk cruise missiles or other payloads, such as large-diameter unmanned underwater vehicles (UUVs).[17]

The four additional launch tubes in the VPM could carry a total of 28 additional Tomahawk cruise missiles (7 per tube),[18] which would increase the total number of torpedo-sized weapons (such as Tomahawks) carried by the Virginia class design from about 37 to about 65—an increase of about 76%.[19] The Navy wants to start building Virginia-class boats with the VPM in FY2019. The Navy's FY2016 five-year shipbuilding plan calls for building one of the two Virginia-class boats to be procured in FY2019, and one of the two Virginia-class boats to be procured in FY2020, with the VPM.

Building Virginia-class boats with the VPM would compensate for a sharp loss in submarine force weapon-carrying capacity that will occur with the retirement in FY2026-FY2028 of the Navy's four Ohio-class cruise missile/special operations forces support submarines (SSGNs).[20] Each SSGN is equipped with 24 large-diameter vertical launch tubes, of which 22 can be used to carry up to 7 Tomahawks each, for a maximum of 154 vertically launched Tomahawks per boat, or 616 vertically launched Tomahawks for the four boats. Twenty-two Virginia-class boats built with VPMs could carry 616 Tomahawks in their VPMs.

The Navy in 2013 estimated that adding the VPM would increase the procurement cost of the Virginia-class design by $350 million in current dollars, or by about 13%.[21]

The joint explanatory statement for the FY2014 DOD Appropriations Act (Division C of H.R. 3547/P.L. 113-76 of January 17, 2014) requires the Navy to submit biannual reports to the congressional defense committees describing the actions the Navy is taking to minimize costs for the VPM.[22] The first such report, dated July 2014, is reprinted in *Appendix C*.[23]

At a February 25, 2015, hearing before the Seapower and Projection Forces subcommittee of the House Armed Services Committee, Sean Stackley, the Assistant Secretary of the Navy for Research, Development, and Acquisition (i.e., the Navy's acquisition executive), stated that the Navy is examining the feasibility of accelerating the procurement of the first VPM-equipped Virginia-class boat from FY2019 to an earlier year.[24]

FY2016 Funding Request

The Navy estimates the combined procurement cost of the two Virginia-class boats requested for procurement in FY2016 at $5,376.9 million or an average of $2,688.4 million each. The boats have received a total of $1,613.5 million in prior-year advance procurement (AP) funding and $416.9 million in prior-year Economic Order Quantity (EOQ) funding. The Navy's proposed FY2016 budget requests the remaining $3,346.4 million needed to complete the boats' estimated combined procurement cost. The Navy's proposed FY2016 budget also requests $1,663.8 million in AP funding and $330.0 million in EOQ funding for Virginia-class boats to be procured in future fiscal years, bringing the total FY2016 funding request for the program (excluding outfitting and post-delivery costs) to $5,340.1 million.

The Navy's proposed FY2016 budget also requests $167.7 million in research and development funding for the Virginia Payload Module (VPM). The funding is contained in Program Element (PE) 0604580N, entitled Virginia Payload Module (VPM), which is line 123 in the Navy's FY2016 research and development account.

Submarine Construction Industrial Base

In addition to GD/EB and NNS, the submarine construction industrial base includes scores of supplier firms, as well as laboratories and research facilities, in numerous states. Much of the total material procured from supplier firms for the construction of submarines comes from single or sole source suppliers. Observers in recent years have expressed concern for the continued survival of many of these firms. For nuclear-propulsion component suppliers, an additional source of stabilizing work is the Navy's nuclear-powered aircraft carrier construction program.[25] In terms of work provided to these firms, a carrier nuclear propulsion plant is roughly equivalent to five submarine propulsion plants.

Much of the design and engineering portion of the submarine construction industrial base is resident at GD/EB. Smaller portions are resident at NNS and some of the component makers.

Several years ago, some observers expressed concern about the Navy's plans for sustaining the design and engineering portion of the submarine construction industrial base. These concerns appear to have receded, in large part because of the Navy's plan to design and procure a next-generation ballistic missile submarine called the Ohio Replacement Program or SSBN(X).[26]

Projected SSN Shortfall

Size and Timing of Shortfall

The Navy's FY2016 30-year SSN procurement plan, if implemented, would not be sufficient to maintain a force of 48 SSNs consistently over the long run. As shown in *Table 2*, the Navy projects under the plan that the SSN force would fall below 48 boats starting in FY2025, reach a minimum of 41 boats in FY2029, and remain below 48 boats through FY2036. Since the Navy plans to retire the four SSGNs by 2028 without procuring any replacements for them, no SSGNs would be available in 2028 and subsequent years to help compensate for a drop in SSN force level below 48 boats.

The projected SSN shortfall was first identified by CRS in 1995 and has been discussed in CRS reports and testimony every year since then.

Table 2. Projected SSN Shortfall
As shown in Navy's FY2016 30-Year (FY2016-FY2045) Shipbuilding Plan

Fiscal year	Annual procurement quantity	Projected number of SSNs	Shortfall relative to 48-boat goal	
			Number of ships	Percent
16	2	53		
17	2	50		
18	2	52		
19	2	50		
20	2	51		
21	1	51		
22	2	48		

Table 2. (Continued)

Fiscal year	Annual procurement quantity	Projected number of SSNs	Shortfall relative to 48-boat goal	
23	2	49		
24	1	48		
25	2	47	-1	-2%
26	1	45	-3	-6%
27	1	44	-4	-8%
28	1	42	-6	-13%
29	1	41	-7	-15%
30	1	42	-6	-13%
31	1	43	-5	-10%
32	1	43	-5	-10%
33	1	44	-4	-8%
34	1	45	-3	-6%
35	1	46	-2	-4%
36	2	47	-1	-2%
37	2	48		
38	2	47	-1	-2%
39	2	47	-1	-2%
40	1	47	-1	-2%
41	2	47	-1	-2%
42	1	49		
43	2	49		
44	1	50		
45	2	50		

Source: Table prepared by CRS based on Navy's FY2016 30-year shipbuilding plan. Percent figures rounded to nearest percent.

2006 Navy Study on Options for Mitigating Projected Shortfall

The Navy in 2006 initiated a study on options for mitigating the projected SSN shortfall. The study was completed in early 2007 and briefed to CRS and the Congressional Budget Office (CBO) on May 22, 2007.[27] At the time of the study, the SSN force was projected to bottom out at 40 boats and then recover to 48 boats by the early 2030s. Principal points in the Navy study (which cite SSN force-level projections as understood at that time) include the following:

- The day-to-day requirement for deployed SSNs is 10.0, meaning that, on average, a total of 10 SSNs are to be deployed on a day-to-day basis.[28]

- The peak projected wartime demand is about 35 SSNs deployed within a certain amount of time. This figure includes both the 10.0 SSNs that are to be deployed on a day-to-day basis and 25 additional SSNs surged from the United States within a certain amount of time.[29]
- Reducing Virginia-class shipyard construction time to 60 months—something that the Navy already plans to do as part of its strategy for meeting the Virginia-class cost-reduction goal (see earlier discussion on cost-reduction goal)—will increase the size of the SSN force by two boats, so that the force would bottom out at 42 boats rather than 40.[30]
- If, in addition to reducing Virginia-class shipyard construction time to 60 months, the Navy also lengthens the service lives of 16 existing SSNs by periods ranging from 3 months to 24 months (with many falling in the range of 9 to 15 months), this would increase the size of the SSN force by another two boats, so that the force would bottom out at 44 boats rather than 40 boats.[31] The total cost of extending the lives of the 16 boats would be roughly $500 million in constant FY2005 dollars.[32]
- The resulting force that bottoms out at 44 boats could meet the 10.0 requirement for day-to-day deployed SSNs throughout the 2020-2033 period if, as an additional option, about 40 SSN deployments occurring in the eight-year period 2025-2032 were lengthened from six months to seven months. These 40 or so lengthened deployments would represent about one-quarter of all the SSN deployments that would take place during the eight-year period.
- The resulting force that bottoms out at 44 boats could not meet the peak projected wartime demand of about 35 SSNs deployed within a certain amount of time. The force could generate a total deployment of 32 SSNs within the time in question—3 boats (or about 8.6%) less than the 35-boat figure. Lengthening SSN deployments from six months to seven months would not improve the force's ability to meet the peak projected wartime demand of about 35 SSNs deployed within a certain amount of time.
- To meet the 35-boat figure, an additional four SSNs beyond those planned by the Navy would need to be procured. Procuring four additional SSNs would permit the resulting 48-boat force to surge an additional three SSNs within the time in question, so that the force could meet the peak projected wartime demand of about 35 SSNs deployed within a certain amount of time.

- Procuring one to four additional SSNs could also reduce the number of seven-month deployments that would be required to meet the 10.0 requirement for day-to-day deployed SSNs during the period 2025-2032. Procuring one additional SSN would reduce the number of seven-month deployments during this period to about 29; procuring two additional SSNs would reduce it to about 17, procuring three additional SSNs would reduce it to about 7, and procuring four additional SSNs would reduce it to 2.

The Navy added a number of caveats to these results, including but not limited to the following:

- The requirement for 10.0 SSNs deployed on a day-to-day basis is a current requirement that could change in the future.
- The peak projected wartime demand of about 35 SSNs deployed within a certain amount of time is an internal Navy figure that reflects recent analyses of potential future wartime requirements for SSNs. Subsequent analyses of this issue could result in a different figure.
- The identification of 19 SSNs as candidates for service life extension reflects current evaluations of the material condition of these boats and projected use rates for their nuclear fuel cores. If the material condition of these boats years from now turns out to be worse than the Navy currently projects, some of them might no longer be suitable for service life extension. In addition, if world conditions over the next several years require these submarines to use up their nuclear fuel cores more quickly than the Navy now projects, then the amounts of time that their service lives might be extended could be reduced partially, to zero, or to less than zero (i.e., the service lives of the boats, rather than being extended, might need to be shortened).
- The analysis does not take into account potential rare events, such as accidents, that might force the removal an SSN from service before the end of its expected service life.[33]
- Seven-month deployments might affect retention rates for submarine personnel.

Issues for Congress

Virginia-Class Procurement Rate More Generally in Coming Years

One potential issue for Congress concerns the Virginia-class procurement rate in coming years, particularly in the context of the SSN shortfall projected for FY2025-FY2036 shown in *Table 2* and the larger debate over future U.S. defense strategy and defense spending.

Mitigating Projected SSN Shortfall

In addition to lengthening SSN deployments to 7 months and extending the service lives of existing SSNs by periods ranging from 3 months to 24 months (see "2006 Navy Study on Options for Mitigating Projected Shortfall" above), options for more fully mitigating the projected SSN shortfall include

- refueling a small number of (perhaps one to five) existing SSNs and extending their service lives by 10 years or more, and
- putting additional Virginia-class boats into the 30-year shipbuilding plan.

It is not clear whether it would be feasible or cost-effective to refuel existing SSNs and extend their service lives by 10 or more years, given factors such as limits on submarine pressure hull life.

Larger Debate on Defense Strategy and Defense Spending

Some observers—particularly those who propose reducing U.S. defense spending as part of an effort to reduce the federal budget deficit—have recommended that the SSN force-level goal be reduced to something less than 48 boats, and/or that Virginia-class procurement be reduced. A June 2010 report from a group called the Sustainable Defense Task Force recommends a Navy of 230 ships, including 37 SSNs,[34] and a September 2010 report from the Cato Institute recommends a Navy of 241 ships, including 40 SSNs.[35] Both reports recommend limiting Virginia-class procurement to one boat per year, as does a September 2010 report from the Center for American Progress.[36] A November 2010 report from a group called the Debt Reduction Task Force recommends "deferring" Virginia-class procurement.[37] The November 2010 draft recommendations of the co-chairs of the Fiscal Commission include recommendations for reducing procurement of certain weapon systems; the Virginia-class program is not among them.

Other observers have recommended that the SSN force-level goal should be increased to something higher than 48 boats, particularly in light of Chinese naval modernization.[38] The July 2010 report of an independent panel that assessed the 2010 Quadrennial Defense Review (QDR)—an assessment that is required by the law governing QDRs (10 U.S.C. 118)— recommends a Navy of 346 ships, including 55 SSNs.[39] An April 2010 report from the Heritage Foundation recommends a Navy of 309 ships, including 55 SSNs.[40]

Factors to consider in assessing whether to maintain, increase, or reduce the SSN force-level goal and/or planned Virginia-class procurement include but are not limited to the federal budget and debt situation, the value of SSNs in defending U.S. interests and implementing U.S. national security strategy, and potential effects on the submarine industrial base.

As discussed earlier, Virginia-class boats scheduled for procurement in FY2014 are covered under an MYP contract for the period FY2014-FY2018. This MYP contract includes the procurement of two Virginia-class boats in FY2016. If fewer than two boats were procured in FY2016, the Navy might need to terminate the MYP contract and pay a cancellation penalty to the contractor.

Accelerating Start of VPM Procurement

Another potential issue for Congress is whether to accelerate the procurement of the first VPMequipped Virginia-class boat from FY2019 to an earlier year. As discussed above, the Navy testified on February 25, 2015, that it is examining the feasibility this option.

LEGISLATIVE ACTIVITY FOR FY2016

FY2016 Funding Request

The Navy estimates the combined procurement cost of the two Virginia-class boats requested for procurement in FY2016 at $5,376.9 million or an average of $2,688.4 million each. The boats have received a total of $1,613.5 million in prior-year advance procurement (AP) funding and $416.9 million in prior-year Economic Order Quantity (EOQ) funding. The Navy's proposed

FY2016 budget requests the remaining $3,346.4 million needed to complete the boats' estimated combined procurement cost. The Navy's proposed FY2016 budget also requests $1,663.8 million in AP funding and $330.0 million in EOQ funding for Virginia-class boats to be procured in future fiscal years, bringing the total FY2016 funding request for the program (excluding outfitting and post-delivery costs) to $5,340.1 million.

The Navy's proposed FY2016 budget also requests $167.7 million in research and development funding for the Virginia Payload Module (VPM). The funding is contained in Program Element (PE) 0604580N, entitled Virginia Payload Module (VPM), which is line 123 in the Navy's FY2016 research and development account.

APPENDIX A. PAST SSN FORCE-LEVEL GOALS

This appendix summarizes attack submarine force-level goals since the Reagan Administration (1981-1989).

The Reagan-era plan for a 600-ship Navy included an objective of achieving and maintaining a force of 100 SSNs.

The George H. W. Bush Administration's proposed Base Force plan of 1991-1992 originally called for a Navy of more than 400 ships, including 80 SSNs.[41] In 1992, however, the SSN goal was reduced to about 55 boats as a result of a 1992 Joint Staff force-level requirement study (updated in 1993) that called for a force of 51 to 67 SSNs, including 10 to 12 with Seawolf-level acoustic quieting, by the year 2012.[42]

The Clinton Administration, as part of its 1993 Bottom-Up Review (BUR) of U.S. defense policy, established a goal of maintaining a Navy of about 346 ships, including 45 to 55 SSNs.[43] The Clinton Administration's 1997 QDR supported a requirement for a Navy of about 305 ships and established a tentative SSN force-level goal of 50 boats, "contingent on a reevaluation of peacetime operational requirements."[44] The Clinton Administration later amended the SSN figure to 55 boats (and therefore a total of about 310 ships).

The reevaluation called for in the 1997 QDR was carried out as part of a Joint Chiefs of Staff (JCS) study on future requirements for SSNs that was completed in December 1999. The study had three main conclusions:

- "that a force structure below 55 SSNs in the 2015 [time frame] and 62 [SSNs] in the 2025 time frame would leave the CINC's [the regional military commandersin-chief] with insufficient capability to respond to urgent crucial demands without gapping other requirements of higher national interest. Additionally, this force structure [55 SSNs in 2015 and 62 in 2025] would be sufficient to meet the modeled war fighting requirements";

- "that to counter the technologically pacing threat would require 18 Virginia class SSNs in the 2015 time frame"; and
- "that 68 SSNs in the 2015 [time frame] and 76 [SSNs] in the 2025 time frame would meet all of the CINCs' and national intelligence community's highest operational and collection requirements."[45]

The conclusions of the 1999 JCS study were mentioned in discussions of required SSN force levels, but the figures of 68 and 76 submarines were not translated into official Department of Defense (DOD) force-level goals.

The George W. Bush Administration's report on the 2001 QDR revalidated the amended requirement from the 1997 QDR for a fleet of about 310 ships, including 55 SSNs. In revalidating this and other U.S. military force-structure goals, the report cautioned that as DOD's "transformation effort matures—and as it produces significantly higher output of military value from each element of the force—DOD will explore additional opportunities to restructure and reorganize the Armed Forces."[46]

DOD and the Navy conducted studies on undersea warfare requirements in 2003-2004. One of the Navy studies—an internal Navy study done in 2004—reportedly recommended reducing the attack submarine force level requirement to as few as 37 boats. The study reportedly recommended homeporting a total of nine attack submarines at Guam and using satellites and unmanned underwater vehicles (UUVs) to perform ISR missions now performed by attack submarines.[47]

In March 2005, the Navy submitted to Congress a report projecting Navy force levels out to FY2035. The report presented two alternatives for FY2035—a 260-ship fleet including 37 SSNs and 4 SSGNs, and a 325-ship fleet including 41 SSNs and 4 SSGNs.[48]

In May 2005, it was reported that a newly completed DOD study on attack submarine requirements called for maintaining a force of 45 to 50 boats.[49]

In February 2006, the Navy proposed to maintain in coming years a fleet of 313 ships, including 48 SSNs. Some of the Navy's ship force-level goals have changed since 2006, and the goals now add up to a desired fleet of 328 ships. The figure of 48 SSNs, however, remains unchanged from 2006.

APPENDIX B. OPTIONS FOR FUNDING SSNS

This appendix presents information on some alternatives for funding SSNs that was originally incorporated into this report during discussions in earlier years on potential options for Virginia-class procurement.

Alternative methods of funding the procurement of SSNs include but are not necessarily limited to the following:

- *two years of advance procurement funding followed by full funding*—the traditional approach, under which there are two years of advance procurement funding for the SSN's long-leadtime components, followed by the remainder of the boat's procurement funding in the year of procurement;
- *one year of advance procurement funding followed by full funding*—one year of advance procurement funding for the SSN's long-leadtime components, followed by the remainder of the boat's procurement funding in the year of procurement;
- *full funding with no advance procurement funding (single-year full funding)*—full funding of the SSN in the year of procurement, with no advance procurement funding in prior years;
- *incremental funding*—partial funding of the SSN in the year of procurement, followed by one or more years of additional funding increments needed to complete the procurement cost of the ship; and
- *advance appropriations*—a form of full funding that can be viewed as a legislatively locked in form of incremental funding.[50]

Navy testimony to Congress in early 2007, when Congress was considering the FY2008 budget, suggested that two years of advance procurement funding are required to fund the procurement of an SSN, and consequently that additional SSNs could not be procured until FY2010 at the earliest.[51] This testimony understated Congress's options regarding the procurement of additional SSNs in the near term. Although SSNs are normally procured with two years of advance procurement funding (which is used primarily for financing long-leadtime nuclear propulsion components), Congress can procure an SSN without prior-year advance procurement funding, or with only one year of advance procurement funding. Consequently, Congress at that time had option of procuring an additional SSN in FY2009 and/or FY2010.

Single-year full funding has been used in the past by Congress to procure nuclear-powered ships for which no prior-year advance procurement funding had been provided. Specifically, Congress used single-year full funding in FY1980 to procure the nuclear-powered aircraft carrier CVN-71, and again in FY1988 to procure the CVNs 74 and 75. In the case of the FY1988 procurement, under the Administration's proposed FY1988 budget, CVNs 74 and 75 were to be procured in FY1990 and FY1993, respectively, and the FY1988 budget was to make the initial advance procurement payment for CVN-74. Congress, in acting on the FY1988 budget, decided to accelerate the procurement of both ships to FY1988, and fully funded the two ships that year at a combined cost of $6.325 billion. The ships entered service in 1995 and 1998, respectively.[52]

The existence in both FY1980 and FY1988 of a spare set of Nimitz-class reactor components was not what made it possible for Congress to fund CVNs 71, 74, and 75 with single-year full funding; it simply permitted the ships to be built more quickly. What made it possible for Congress to fund the carriers with single-year full funding was Congress's constitutional authority to appropriate funding for that purpose.

Procuring an SSN with one year of advance procurement funding or no advance procurement funding would not materially change the way the SSN would be built—the process would still encompass about two years of advance work on long-leadtime components, and an additional six years or so of construction work on the ship itself. The outlay rate for the SSN could be slower, as outlays for construction of the ship itself would begin one or two years later than normal.

Congress in the past has procured certain ships in the knowledge that those ships would not begin construction for some time and consequently would take longer to enter service than a ship of that kind would normally require. When Congress procured two nuclear-powered aircraft carriers (CVNs 72 and 73) in FY1983, and another two (CVNs 74 and 75) in FY1988, it did so in both cases in the knowledge that the second ship in each case would not begin construction until some time after the first.

APPENDIX C. JULY 2014 NAVY REPORT TO CONGRESS ON VIRGINIA PAYLOAD MODULE (VPM)

The joint explanatory statement for the FY2014 DOD Appropriations Act (Division C of H.R. 3547/P.L. 113-76 of January 17, 2014) requires the Navy to submit biannual reports to the congressional defense committees describing the actions the Navy is taking to minimize costs for the VPM.[53] This appendix reprints the first of these reports, which is dated July 2014.[54]

REPORT TO CONGRESS

VIRGINIA (SSN774) ATTACK CLASS SUBMARINE

Cost Containment Strategy
for the
Block V VIRGINIA Payload Module (VPM) Design

July 2014

IN COMPLIANCE WITH THE JOINT EXPLANATORY STATEMENT ACCOMPANYING
THE
CONSOLIDATED APPROPRIATIONS ACT, 2014

PREPARED BY
PROGRAM EXECUTIVE OFFICE, SUBMARINES
614 SICARD STREET
WASHINGTON NAVY YARD, DC 20376

The estimated cost of this report or study for the
Department of Defense is approximately $10,000 for the 2014 Fiscal Year.
This includes $0 in expenses and $10,000 in DoD labor.
Generated on 2014Jul24 RefID: D-1DF2309

Table of Contents

Reporting Requirement

Division C of the Joint Explanatory Statement accompanying the Consolidated Appropriations Act, 2014 (Public Law 113-76), directs the Secretary of the Navy to create a separate budget line item to enable additional congressional oversight and increase transparency into the costs of the VIRGINIA Payload Module (VPM). Furthermore, Congress directed the Secretary to submit a bi-annual report to the congressional defense committees describing the actions the Navy is taking to minimize costs. The agreement fences $20 million until the first bi-annual report is provided to the congressional defense committees.

Executive Summary

In the mid-2020s, the Navy's four guided missile submarines (SSGNs) will begin to decommission. These SSGNs provide the Navy and the Nation with unmatched undersea conventional strike capability and capacity, with each SSGN carrying up to 154 Tomahawk land attack cruise missiles. The Navy's current fleet of attack submarines (SSNs) can carry 12 Tomahawks each. The loss of the SSGNs will result in an over 60 percent drop in undersea strike capacity.

The Department of Defense's Office of Cost and Program Evaluation (CAPE) conducted a review of the potential undersea strike alternatives to determine the optimal materiel solution to recapitalize the SSGNs' strike capacity. CAPE certified to the Office of the Under Secretary of Defense (Acquisition, Technology and Logistics) (AT&L) that the Navy studies in conjunction with CAPE's independent review and the Naval Sea Systems Command's (NAVSEA) Cost Engineering and Industrial Analysis's (05C) cost estimate met the requirements of an Analysis of Alternatives (AoA), and CAPE did not recommend performing an AoA for undersea strike. The review determined that the VIRGINIA Payload Module (VPM), a hull insert with four large-diameter tubes inserted aft of the sail, each tube capable of carrying seven Tomahawks, represented the best materiel solution to mitigate the loss of undersea strike capacity given near-term budget constraints. To minimize cost, schedule, and technical risks, VPM will reuse operationally proven systems and will not require the development of any new technology. For example, the missile tubes that will be used in VPM are nearly identical to the multiple all-up-round canister (MAC) tubes that are currently deployed on the SSGNs.

In December 2013, the Joint Requirement Oversight Council (JROC) approved the Capability Development Document (CDD) establishing the requirements and Key Performance Parameters (KPPs) for VPM. The CDD set clear KPPs for cost, schedule, and strike capacity. By placing cost on equal footing as capability, the CDD ensures the Navy will leverage its best practices and lessons learned from previous submarine research and development, acquisition, and modernization efforts to deliver the required capability within the strict cost targets.

Alteration to the design of any weapon system in full rate production has the potential to introduce justifiable concern associated with the possible erosion of program cost performance and production. The Navy recognizes these risks as they apply to implementation of VPM during Block V construction and intends to employ a full range of management techniques to mitigate them, commencing early in the design phase. The Navy has a proven record of developing and executing similarly scaled efforts such as the Block III design for affordability effort including the redesigned bow. These techniques are well established and embedded in the current submarine acquisition community culture, developed during "NSSN" [the New Attack Submarine Program – the precursor to the VIRGINIA Class] program inception and evolved through the successful VIRGINIA Class Block IV construction contract award.

The Navy's disciplined engineering and acquisition management approach for VPM, in conjunction with treating cost and capability as equally important requirements, will minimize the potential for cost performance degradation and program disruption. The key actions the Navy is taking to minimize costs are: continue proven management techniques used from program inception through Block IV award; implementation of Integrated Product and Process Development (IPPD) in conjunction with execution of existing build plans; ensure stable requirements; high design completion at construction start; risk mitigation; and cost reporting.

4

1. Background

The VIRGINIA Class Submarine Program was the first major defense program to implement the tenets of the October 1994 Under Secretary of Defense for Acquisition and Technology memorandum, "Implementation of Integrated Product and Process Development (IPPD) in DoD Acquisition Programs." The VIRGINIA Class program has continuously implemented the use of Commercial Off-the-Shelf (COTS) components, open systems standards, acquisition streamlining, total ownership cost (TOC) driven decision making, Lean 6 Sigma assessments of all processes, and recent should cost/will cost and Better Buying Power initiatives to improve the program as it has matured.

1.1 Block I – IPPD Design/Build Genesis (SSNs 774–777)

From inception, the VIRGINIA Class Submarine Program was strikingly different from past fast attack programs, in part due to advances in technology, but mostly due to revolutionary changes in the design/build, business, and acquisition processes. The Navy, General Dynamics Electric Boat (GDEB) and their major subcontractor, Huntington Ingalls Industries – Newport News Shipbuilding (HII-NNS), embraced the IPPD concept and established multi-disciplined teams to collaboratively design and build the submarine. Inherent in the definition of IPPD, both products and processes derived benefit from structured and hierarchical integration of the cross-functional teams. The IPPD approach holistically linked operational performance, construction techniques, test methods, and life-cycle supportability into an up-front "single-pass" design effort. IPPD enabled the shipbuilder to expand the use of modular construction and off-hull module assembly techniques beyond that of previous submarine programs and erect the entire submarine from 10 major sections. While the IPPD approach was exceedingly effective, the introduction of a new, sophisticated Computer Aided Three-dimensional Interactive Application (CATIA) also greatly enhanced the design/build process and programmatic business efficiency. The CATIA software design tool replaced traditional drawings and hand crafted wooden models with 3-D manipulative color graphics dispersed to Integrated Product Team members to facilitate timely and efficient, visual design collaboration. CATIA also established the single shipbuilding construction and procurement database, linking design with production and business operations. CATIA also provided a higher fidelity design release forecast which in turn supported the establishment of a more accurate budget baseline from which to conduct cost analysis.

1.2 Block II – Continuous Improvement via Capital Expenditure (SSNs 778–783)

As the program began construction on the Block II submarines, the Navy set about to improve construction efficiencies beginning with USS *New Hampshire* (SSN 778), the first submarine in the Block II contract. Recognizing construction span time reduction held the most immediate promise for lowering cost and accelerating delivery of the warships, focus was directed at determining what could be done to improve industrial efficiency without compromise to quality or performance. Teaming for success, the Navy and shipbuilders agreed that facility investment was needed, and a strategy to incorporate an innovative Capital Expenditure (CAPEX) incentive clause was devised and incorporated in the Block II contract. Of the 10 Block II CAPEX funded projects, the transportation system upgrades provided the most visible evidence of reduced span time by allowing a shift from the Block I 10 module build plan to a plan entailing only four "super" modules to undergo final assembly at the delivery shipyard. Block II CAPEX projects have produced a seven to one return on investment.

1.3 Block III – Design for Affordability (DFA) (SSNs 784–791)

The VIRGINIA Class cost reduction program began in earnest in late 2005, when the Chief of Naval Operations (CNO) issued a challenge to the VIRGINIA Class Program to reduce the acquisition cost of each submarine to $2 billion (in FY 2005 dollars) by 2012 as a condition of increasing the procurement rate from one to two submarines per year. This challenge represented a 20 percent decrease in unit cost. The CNO issued the challenge to support the acquisition of VIRGINIA Class submarines within the Navy shipbuilding budget, and make the necessary increase in attack submarine production to support national force levels. At the time, cutting 20 percent of the unit cost of a submarine with a mature design in serial production, without removing capability, was an unprecedented task.

General Dynamics Electric Boat and the Navy developed an integrated cost reduction strategy focusing on three areas – Design for Affordability (DFA), construction performance, and acquisition/procurement strategy. Implementation of all three elements resulted in a savings of $400 million per ship and a reduction in construction span by 30 percent (from 84 to 60 months). In recognition of these collaborative achievements, the VIRGINIA Class Program received the 2008 David Packard Excellence in Acquisition Award for "embracing the principles of acquisition reform since its initiation," and for having significantly reduced total program costs.

1.4 Block IV – Reduced Total Ownership Cost (RTOC) (SSNs 792–801)

Having optimized the construction process via targeted capital investment and DFA, the program concentrated on creating more operational value from each submarine by increasing the time between major maintenance availabilities. The goal was to alter the established life cycle maintenance plan from 72-month operating cycles, with 14 deployments and four major depot availabilities, to 96-month operating cycles, with 15 deployments and only three major depot availabilities. The challenge once again was to identify which design changes offered the highest Reduction of Total Ownership Cost (RTOC) return on investment from a limited design budget – assessing maintenance drivers and factors that determine the aggregate operating cycle. By eliminating one depot availability per hull, the program will avoid approximately $120 million (FY 2010 dollars) in Operating and Support costs per submarine. By enabling an additional deployment from each subsequent Block IV and beyond hull, an operational availability equivalent to one submarine will be realized following delivery of SSN 805.

2.0 Block V –VPM Concept Origination

The VPM concept was introduced to address the eventual loss of submarine guided missile (SSGN) strike capabilities in the mid-2020s when the Navy's four SSGNs retire, reducing Navy-wide undersea strike volume by almost two-thirds. The SSGNs' retirement also coincides with a historically low attack class submarine force structure.

In a 2013 review of undersea strike alternatives conducted by CAPE, VPM was identified to be the optimal materiel solution to recapitalize undersea strike without substantially changing a mature and stable submarine design. CAPE certified to AT&L that the review met the requirements of an AoA, and an AoA was not required. VIRGINIA Class submarines with VPM would retain all existing mission capability, while providing approximately 94 percent of the current undersea strike volume.

In December 2013, the JROC approved the CDD establishing the requirements and KPPs for VPM. The CDD sets clear KPPs for strike capacity, schedule, and cost. The strike KPP increases the missile capacity from 12 to 40. For schedule, the VPM's Initial Operating Capability (IOC) threshold and objective dates are no later than 2nd quarter FY 2028 and no later than 4th quarter FY 2026, respectively.

The cost KPP includes criteria for design, lead ship, and follow ship thresholds and objectives requiring a disciplined approach to balance capabilities within the established cost parameters. Based on the NAVSEA 05C current estimate, the VPM cost estimate is below the CDD's cost objectives.

Cost - CY10$ ($M)

	Threshold	Objective	Current Est.
NRE:	800	750	744
Lead Ship:	475	425	423
Follow on Ships	350	325	318

Cost - TY$ ($M)

	Threshold	Objective	Current Est.
NRE:	994	931	924
Lead Ship:	633	567	564
Follow on Ships	567	527	515

Note: CDD Cost values are for 20 VPM modules and start of construction in FY19

The Navy/Industry team is focused on controlling VPM program costs, while minimizing baseline ship impacts, and maintaining the established VIRGINIA Class build plan cadence. As a result of the VIRGINIA Class modular design, inherent design features make the insertion of a hull section less of an impact on the build plan. The VPM design is modeled after other successful VIRGINIA Class programs, which have lowered costs through a proven cost reduction framework.

3.0 FY 2014 VPM Design Funding and Cost Control Management Requirements

The Consolidated Appropriations Act, 2014 (Public Law 113-76) appropriated $59.1 million for the development of VPM. Division C of the Joint Explanatory Statement accompanying the Consolidated Appropriations Act, 2014, directed the creation of a separate budget line item to enable additional congressional oversight and increase transparency into the costs of the VPM. The Navy established *Navy PE: 0604580N VIRGINIA Payload Module (VPM)* to fulfill this requirement. The Joint Explanatory Statement also stipulated the withholding of $20 million in funding until the first submission of a bi-annual report to the congressional defense committees describing the actions the Navy plans to take to minimize costs. The following sections of this report are intended to fulfill this requirement.

4.0 Cost Containment Strategy for the Block V VPM Design

The strategy to design and seamlessly insert VPM into the construction sequence within the established budget is to employ the full spectrum of proven management techniques used from program inception through Block IV contract award. Specifically:

- Incorporate key tenets of the USD(AT&L) Better Buying Power 2.0 approach to defense acquisition such as affordability targets and innovative contract incentives.
- Apply overarching IPPD practices and implement design/build teams (Block I and III lessons learned).

7

- Identify capital investment opportunities with high return on investment potential (Block II and III lessons learned).
- Develop design focused on affordability (Block III lessons learned) and life cycle maintenance costs (Block IV lessons learned).
- Explore and establish ship and component level acquisition strategies to yield a higher confidence/lower cost construction cost (Block III and IV lessons learned).
- Utilize an incentive structure that specifically details required cost reductions in design, construction, and operations and support.

These techniques have guided the VIRGINIA Class Program and will be used throughout the VPM effort.

4.1 Implementation of IPPD in conjunction with execution of existing build plan

The IPPD approach that was utilized as part of the successful Block III bow redesign effort provided the program with the experience and the strategy that can be leveraged for VPM during ongoing production. This will ensure the VPM design is strategically coordinated with construction and will not disrupt the established four-module build plan or construction cadence. This, in turn, requires an increase in the Advance Procurement funding profile for Block V to enable the completion of VPM during the fabrication and assembly phase at the same time as the other module components. A detailed Integrated Master Schedule (IMS) and Module Build Plan will be completed in December 2014, providing the comprehensive IPPD roadmap to minimize baseline ship impacts and maintain the established VIRGINIA Class construction cadence. In addition, the design team will evaluate capital investment opportunities to lower construction costs.

4.2 Stable requirements

The CDD sets clear KPPs for cost, strike capacity, and schedule based on stable requirements. These KPPs promote stability in the Program, providing the Navy and Shipbuilders with fixed, tangible, and measurable objectives. By placing cost on equal footing as capability, the CDD ensures the Navy will leverage its best practices and lessons learned from previous submarine research and development, acquisition, and modernization efforts to deliver the required capability within the strict cost targets. The ship specification process will further define the requirements in strict accordance with the KPPs.

4.3 Design completion

The current VPM design concept does not require the development of any new technology to satisfy the CDD requirements. By relying on proven operational systems, the Navy avoids the unnecessary risk new technology poses. Similarly, "like" systems and components already utilized or proven elsewhere in the submarine enterprise will be leveraged, scaled, or reused to an extensive degree. The most obvious example of this strategy pertains to replication of the tubes and sealing of the launch control electronics from the bow of the Block III design. The collective sum of the re-use strategy tied to the VIRGINIA Payload Tubes (VPTs), Submarine Warfare Federated Tactical System (SWFTS) combat system, Ship Service Hydraulic Plant, Electronic Auxiliary Fresh Water Plant, and other Hull, Mechanical and Electrical subsystems results in a high Technology Readiness Level (TRL) for the VPM effort. This equates to an achievable goal of having the VPM design 80 percent complete prior to construction start, adding confidence to completing the design within budget and minimizing construction costs.

4.4 Risk mitigation

The VPM cost reduction program will employ a low-risk technical approach, with a goal of having the VPM design 80 percent complete prior to construction start. This will ensure that design errors do not create issues during the construction phase, thereby avoiding unforeseen costs later in the program. With no new technology and significant design and component reuse, the VPM design has a high TRL, thus low risk to the shipbuilder. The program will continue to evaluate and mitigate construction and design risk. For example, the program will benefit when the land based VPT test site is completed at Naval Undersea Warfare Center (NUWC) Newport this fall. Manufactured at Quonset Point and installed by Electric Boat, this collaborative Navy/shipbuilder test facility will support early electronic testing to mitigate VPM risk, and lower shipbuilding construction risk.

The shipbuilding industrial base is well positioned to simultaneously design both VPM and OHIO Replacement as the completion of the VIRGINIA Block III and Moored Training Ship design efforts allow for sufficient General Dynamics Electric Boat (GDEB) resources to support both designs.

The VIRGINIA Program is collaborating with the OHIO Replacement Program to ensure commonality among select ship components and design features which will benefit the acquisition and life-cycle costs for both programs. Where possible, the programs will utilize common equipment designs such as Ship Control System hardware, and Command, Control, Communications, and Intelligence (C3I) systems. The two programs will utilize best manufacturing processes and practices to ensure cost savings across both classes.

4.5 Cost reporting

The VPM program will continue to use the established best practices that enabled previous cost reduction. The program has an effective and established metrics/performance measurement system to manage cost, schedule and risk. A key and essential factor governing effectiveness is the accuracy of the underlying work scope comprising the budget baseline being tracked. The CATIA design application has remained in use since Block I and provides this essential fidelity. Cost analysis data, combined shipbuilder and Navy estimates at completion (EACs), formal risk management program outputs, and quarterly design reviews will all be utilized to assess the VPM program health. To promote specific transparency into cost, as directed, a separate Research, Development, Test and Evaluation (RDT&E) Program Element (PE: 0604580N) was developed for VPM funding. This new PE is reflected in the 2015 budget submission to Congress and ensures VPM costs are separate and distinct from the program's overall RDT&E budget. Consistent with the program's history of monitoring cost, cost estimates for VPM design will be reviewed quarterly and refined by the VPM design team and the program has developed action plans (based on estimates of cost-at-completion) to track cost reporting.

5.0 Conclusion

This report provides a baseline understanding of VPM and the cost reduction and containment strategies employed by the Navy throughout the VIRGINIA Class Program to include the early efforts on VPM.

Subsequent bi-annual reports will provide additional specific metrics for VPM as its acquisition, design, and construction strategies are developed and refined. Products such as design curves, manning ramp-up plans, design drawings, and progress on ship specifications will be provided with future reports as they become available.

End Notes

1 See CRS Report R41129, Navy Ohio Replacement (SSBN[X]) Ballistic Missile Submarine Program: Background and Issues for Congress, by Ronald O'Rourke.

[2] In U.S. Navy submarine designations, SS stands for submarine, N stands for nuclear-powered, B stands for ballistic missile, and G stands for guided missile (such as a cruise missile). Submarines can be powered by either nuclear reactors or non-nuclear power sources such as diesel engines or fuel cells. All U.S. Navy submarines are nuclear-powered. A submarine's use of nuclear or non-nuclear power as its energy source is not an indication of whether it is armed with nuclear weapons—a nuclear-powered submarine can lack nuclear weapons, and a non-nuclear-powered submarine can be armed with nuclear weapons.

3 The SSBNs' basic mission is to remain hidden at sea with their nuclear-armed submarine-launched ballistic missiles (SLBMs) and thereby deter a strategic nuclear attack on the United States. The Navy's SSBNs are discussed in CRS Report R41129, Navy Ohio Replacement (SSBN[X]) Ballistic Missile Submarine Program: Background and Issues for Congress, by Ronald O'Rourke, and CRS Report RL31623, U.S. Nuclear Weapons: Changes in Policy and Force Structure, by Amy F. Woolf.

[4] The Navy's four SSGNs are former Trident SSBNs that have been converted (i.e., modified) to carry Tomahawk cruise missiles and SOF rather than SLBMs. Although the SSGNs differ somewhat from SSNs in terms of mission orientation (with the SSGNs being strongly oriented toward Tomahawk strikes and SOF support, while the SSNs are more general-purpose in orientation), SSGNs can perform other submarine missions and are sometimes included in counts of the projected total number of Navy attack submarines. The Navy's SSGNs are discussed in CRS Report RS21007, Navy Trident Submarine Conversion (SSGN) Program: Background and Issues for Congress, by Ronald O'Rourke.

[5] For an account of certain U.S. submarine surveillance and intelligence-collection operations during the Cold War, see Sherry Sontag and Christopher Drew with Annette Lawrence Drew, Blind Man's Bluff (New York: Public Affairs, 1998).

[6] For additional information on Navy force-level goals, see CRS Report RL32665, Navy Force Structure and Shipbuilding Plans: Background and Issues for Congress, by Ronald O'Rourke.

[7] Los Angeles-class boats have a beam (i.e., diameter) of 33 feet and a submerged displacement of about 7,150 tons. Seawolf-class boats have a beam of 40 feet. SSN-21 and SSN-22 have a submerged displacement of about 9,150 tons.

[8] SSN-23 is 100 feet longer than SSN-21 and SSN-22 and has a submerged displacement of 12,158 tons.

[9] Virginia-class boats have a beam of 34 feet and a submerged displacement of 7,800 tons.

[10] For a discussion of MYP contracting, see CRS Report R41909, Multiyear Procurement (MYP) and Block Buy Contracting in Defense Acquisition: Background and Issues for Congress, by Ronald O'Rourke and Moshe Schwartz.

[11] For a discussion of block buy contracting, see CRS Report R41909, Multiyear Procurement (MYP) and Block Buy Contracting in Defense Acquisition: Background and Issues for Congress, by Ronald O'Rourke and Moshe Schwartz.

[12] GD/EB and NNS are the only two shipyards in the country capable of building nuclear-powered ships. GD/EB builds submarines only, while NNS also builds nuclear-powered aircraft carriers and is capable of building other types of surface ships.

[13] The joint production arrangement is a departure from prior U.S. submarine construction practices, under which complete submarines were built in individual yards. The joint production arrangement is the product of a debate over the Virginia-class acquisition strategy within Congress, and between Congress and the Department of Defense (DOD), that occurred in 1995-1997 (i.e., during the markup of the FY1996-FY1998 defense budgets). The goal of the arrangement is to keep both GD/EB and NNS involved in building

nuclear-powered submarines, and thereby maintain two U.S. shipyards capable of building nuclear-powered submarines, while minimizing the cost penalties of using two yards rather than one to build a submarine design that is being procured at a relatively low annual rate. The joint production agreement cannot be changed without the agreement of both GD/EB and NNS.

[14] The Navy says that, in constant FY2005 dollars, about $200 million of the $400 million in the sought-after cost reductions was accomplished simply through the improved economies of scale (e.g., better spreading of shipyard fixed costs and improved learning rates) of producing two submarines per year rather than one per year. The remaining $200 million in sought-after cost reductions, the Navy says, was accomplished through changes in the ship's design (which will contribute roughly $100 million toward the cost-reduction goal) and changes in the shipyard production process (which will contribute the remaining $100 million or so toward the goal). Some of the design changes are being introduced to Virginia-class boats procured prior to FY2012, but the Navy said the full set of design changes would not be ready for implementation until the FY2012 procurement. Changes in the shipyard production process are aimed in large part at reducing the total shipyard construction time of a Virginia-class submarine from 72 months to 60 months. (If the ship spends less total time in the shipyard being built, its construction cost will incorporate a smaller amount of shipyard fixed overhead costs.) The principal change involved in reducing shipyard construction time to 60 months involves increasing the size of the modules that form each submarine, so that each submarine can be built out of a smaller number of modules. For detailed discussions of the Virginia-class cost-reduction effort, see David C. Johnson et al., "Managing Change on Complex Programs: VIRGINIA Class Cost Reduction," Naval Engineers Journal, No. 4, 2009: 79-94; and John D. Butler, "The Sweet Smell of Acquisition Success," U.S. Naval Institute Proceedings, June 2011: 22-28.

[15] "Navy Selects Virginia Payload Module Design Concept," USNI News (http://news.usni.org), November 4, 2013.

[16] Christopher P. Cavas, "Innovations, No-Shows At Sea-Air-Space Exhibition," Defense News, April 18, 2011: 4. See also Christopher P. Cavas, "U.S. Navy Eyes Dual-Mission Sub," Defense News, October 17, 2011; and Lee Hudson, "New Virginia-Class Payload Module May Replace SSGN Capability," Inside the Navy, October 24, 2011.

[17] For an illustration of the VPM, see http://www.gdeb.com/news/advertising/images/VPM_ad/VPM.pdf, which was accessed by CRS on March 1, 2012.

[18] Michael J. Conner, "Investing in the Undersea Future," U.S. Naval Institute Proceedings, June 2011: 16-20.

[19] A Virginia-class SSN can carry about 25 Tomahawks or other torpedo-sized weapons in its four horizontal torpedo tubes and associated torpedo room, and an additional 12 Tomahawk cruise missiles in its bow-mounted vertical lunch tubes, for a total of about 37 torpedo-sized weapons. Another 28 Tomahawks in four mid-body vertical tubes would increase that total by about 76%.

[20] Michael J. Conner, "Investing in the Undersea Future," U.S. Naval Institute Proceedings, June 2011: 16-20.

[21] Lee Hudson, "Virginia Payload Module Cost Estimate Down To $350 Million Apiece," Inside the Navy, July 22, 2013. Previously, the Navy had testified that adding the VPM would increase the procurement cost of the Virginia-class design by $360 million to $380 million in current dollars. (Source: Spoken testimony of Sean Stackley, the Assistant Secretary of the Navy for Research, Development, and Acquisition [i.e., the Navy's acquisition executive], at a May 8, 2013, hearing on Navy shipbuilding programs before the Seapower

subcommittee of the Senate Armed Services Committee, as shown in the transcript for the hearing. See also Olga Belogolova, "Navy Officials Lay Out Fragility Of Shipbuilding Budget To Congress," Inside the Navy, May 10, 2013.) Prior to that, the Navy reportedly had estimated that adding the VPM would increase the procurement cost of the Virginia-class design by $400 million to $500 million. (Christopher P. Cavas, "U.S. Navy Eyes Dual-Mission Sub," Defense News, October 17, 2011; see also Michael J. Conner, "Investing in the Undersea Future," U.S. Naval Institute Proceedings, June 2011: 16-20.)

[22] See PDF page 239 of 351 of the joint explanatory statement for Division C of H.R. 3547.

[23] For an article discussing the navy's report, see Lee Hudson, "Stackley Outlines Virginia Payload Module Cost Strategy For Congress," Inside the Navy, November 3, 2014.

[24] Source: Spoken testimony of Assistant Secretary Stackley, as reflected in transcript of hearing.

[25] For more on this program, see CRS Report RS20643, Navy Ford (CVN-78) Class Aircraft Carrier Program: Background and Issues for Congress, by Ronald O'Rourke.

[26] For more on the SBN(X) program, see CRS Report R41129, Navy Ohio Replacement (SSBN[X]) Ballistic Missile Submarine Program: Background and Issues for Congress, by Ronald O'Rourke.

[27] Navy briefing entitled, "SSN Force Structure, 2020-2033," presented to CRS and CBO on May 22, 2007.

[28] The requirement for 10.0 deployed SSNs, the Navy stated in the briefing, was the current requirement at the time the study was conducted.

[29] The peak projected wartime demand of about 35 SSNs deployed within a certain amount of time, the Navy stated, is an internal Navy figure that reflects several studies of potential wartime requirements for SSNs. The Navy stated that these other studies calculated various figures for the number of SSNs that would be required, and that the figure of 35 SSNs deployed within a certain amount of time was chosen because it was representative of the results of these other studies.

[30] If shipyard construction time is reduced from 72 months to 60 months, the result would be a one-year acceleration in the delivery of all boats procured on or after a certain date. In a program in which boats are being procured at a rate of two per year, accelerating by one year the deliveries of all boats procured on or after a certain date will produce a onetime benefit of a single year in which four boats will be delivered to the Navy, rather than two. In the case of the Virginia-class program, this year might be around 2017. As mentioned earlier in the discussion of the Virginia-class cost-reduction goal, the Navy believes that the goal of reducing Virginia-class shipyard construction time is a medium-risk goal. If it turns out that shipyard construction time is reduced to 66 months rather than 60 months (i.e., is reduced by 6 months rather than 12 months), the size of the SSN force would increase by one boat rather than two, and the force would bottom out at 41 boats rather than 42.

[31] The Navy study identified 19 existing SSNs whose service lives currently appear to be extendable by periods of 1 to 24 months. The previous option of reducing Virginia-class shipyard construction time to 60 months, the Navy concluded, would make moot the option of extending the service lives of the three oldest boats in this group of 19, leaving 16 whose service lives would be considered for extension.

[32] The Navy stated that the rough, order-of-magnitude (ROM) cost of extending the lives of 19 SSNs would be $595 million in constant FY2005 dollars, and that the cost of extending the lives of 16 SSNs would be roughly proportional.

[33] In January 2005, the Los Angeles-class SSN San Francisco (SSN-711) was significantly damaged in a collision with an undersea mountain near Guam. The ship was repaired in part by transplanting onto it the bow section of the deactivated sister ship Honolulu (SSN-718).

(See, for example, Associated Press, "Damaged Submarine To Get Nose Transplant," Seattle Post-Intelligencer, June 26, 2006.) Prior to the decision to repair the San Francisco, the Navy considered the option of removing it from service. (See, for example, William H. McMichael, "Sub May Not Be Worth Saving, Analyst Says," Navy Times, February 28, 2005; Gene Park, "Sub Repair Bill: $11M," Pacific Sunday News (Guam), May 8, 2005.)

[34] Debt, Deficits, and Defense, A Way Forward[:] Report of the Sustainable Defense Task Force, June 11, 2010, pp. 19-20, 31.

[35] Benjamin H. Friedman and Christopher Preble, Budgetary Savings from Military Restraint, Washington, Cato Institute, September 23, 2010 (Policy Analysis No. 667), p. 9.

[36] Lawrence J. Korb and Laura Conley, Strong and Sustainable[:] How to Reduce Military Spending While Keeping Our Nation Safe, Center for American Progress, September 2010, pp. 19-20.

[37] Debt Reduction Task Force, Restoring America's Future[:] Reviving the Economy, Cutting Spending and Debt, and Creating a Simple, Pro-Growth Tax System, November 2010, p. 103.

[38] For further discussion of China's naval modernization effort, see CRS Report RL33153, China Naval Modernization: Implications for U.S. Navy Capabilities—Background and Issues for Congress, by Ronald O'Rourke.

[39] Stephen J. Hadley and William J. Perry, co-chairmen, et al., The QDR in Perspective: Meeting America's National Security Needs In the 21st Century, The Final Report of the Quadrennial Defense Review Independent Panel, Washington, 2010, Figure 3-2 on page 58.

[40] A Strong National Defense[:] The Armed Forces America Needs and What They Will Cost, Heritage Foundation, April 5, 2011, pp. 25-26.

[41] For the 80-SSN figure, see Statement of Vice Admiral Roger F. Bacon, U.S. Navy, Assistant Chief of Naval Operations (Undersea Warfare) in U.S. Congress, House Armed Services Committee, Subcommittee on Seapower and Strategic and Critical Materials, Submarine Programs, March 20, 1991, pp. 10-11, or Statement of Rear Admiral Raymond G. Jones, Jr., U.S. Navy, Deputy Assistant Chief of Naval Operations (Undersea Warfare), in U.S. Congress, Senate Armed Services Committee, Subcommittee on Projection Forces and Regional Defense, Submarine Programs, June 7, 1991, pp. 10-11.

[42] See Richard W. Mies, "Remarks to the NSL Annual Symposium," Submarine Review, July 1997, p. 35; "Navy Sub Community Pushes for More Subs than Bottom-Up Review Allowed," Inside the Navy, November 7, 1994, pp. 1, 8-9; Attack Submarines in the Post-Cold War Era: The Issues Facing Policymakers, op. cit., p. 14; Robert Holzer, "Pentagon Urges Navy to Reduce Attack Sub Fleet to 50," Defense News, March 15-21, 1993, p. 10; Barbara Nagy, " Size of Sub Force Next Policy Battle," New London Day, July 20, 1992, pp. A1, A8.

[43] Secretary of Defense Les Aspin, U.S. Department of Defense, Report on the Bottom-Up Review, October 1993, pp. 55-57.

[44] Secretary of Defense William S. Cohen, U.S. Department of Defense, Report of the Quadrennial Defense Review, May 1997, pp. 29, 30, 47.

[45] Department of Navy point paper dated February 7, 2000. Reprinted in Inside the Navy, February 14, 2000, p. 5.

[46] U.S. Department of Defense, Quadrennial Defense Review, September 2001, p. 23.

[47] Bryan Bender, "Navy Eyes Cutting Submarine Force," Boston Globe, May 12, 2004, p. 1; Lolita C. Baldor, "Study Recommends Cutting Submarine Fleet," NavyTimes.com, May 13, 2004.

[48] U.S. Department of the Navy, An Interim Report to Congress on Annual Long-Range Plan for the Construction of Naval Vessels for FY 2006. The report was delivered to the House and Senate Armed Services and Appropriations Committees on March 23, 2005.

[49] Robert A. Hamilton, "Delegation Calls Report on Sub Needs Encouraging," The Day (New London, CT), May 27, 2005; Jesse Hamilton, "Delegation to Get Details on Sub Report," Hartford (CT) Courant, May 26, 2005.

[50] For additional discussion of these funding approaches, see CRS Report RL32776, Navy Ship Procurement: Alternative Funding Approaches—Background and Options for Congress, by Ronald O'Rourke.

[51] For example, at a March 1, 2007, hearing before the House Armed Services Committee on the FY2008 Department of the Navy budget request, Representative Taylor asked which additional ships the Navy might want to procure in FY2008, should additional funding be made available for that purpose. In response, Secretary of the Navy Donald Winter stated in part: "The Virginia-class submarines require us to start with a two-year advanced procurement, to be able to provide for the nuclear power plant that supports them. So we would need to start two years in advance. What that says is, if we were able to start in '08 with advanced procurement, we could accelerate, potentially, the two a year to 2010." (Source: Transcript of hearing.) Navy officials made similar statements before the same subcommittee on March 8, 2007, and before the Senate Armed Services Committee on March 29, 2007.

[52] In both FY1988 and FY1980, the Navy had a spare set of Nimitz (CVN-68) class nuclear propulsion components in inventory. The existence of a spare set of components permitted the carriers to be built more quickly than would have otherwise been the case, but it is not what made the single-year full funding of these carriers possible. What made it possible was Congress's authority to appropriate funds for the purpose.

[53] See PDF page 239 of 351 of the joint explanatory statement for Division C of H.R. 3547.

[54] The report was posted at InsideDefense.com (subscription required) on November 13, 2014.

In: Navy Submarines
Editor: Terry Coleman

ISBN: 978-1-63483-236-6

Chapter 2

NAVY OHIO REPLACEMENT (SSBN[X]) BALLISTIC MISSILE SUBMARINE PROGRAM: BACKGROUND AND ISSUES FOR CONGRESS*

Ronald O'Rourke

SUMMARY

The Navy's proposed FY2016 budget requests $1,390.7 million for continued research and development work on the Ohio replacement program (ORP), a program to design and build a new class of 12 ballistic missile submarines (SSBNs) to replace the Navy's current force of 14 Ohio-class SSBNs. The Ohio replacement program is also known as the SSBN(X) program. The Navy wants to procure the first Ohio replacement boat in FY2021, with advance procurement (AP) funding starting in FY2017. The Navy has identified the Ohio replacement program as its top priority program.

The Navy in June 2014 estimated the procurement cost of the lead ship in the program at $12.4 billion in constant 2014 dollars, including $4.8 billion in detailed design and nonrecurring engineering (DD/NRE) costs for the entire class, and $7.6 billion in construction costs for the ship itself. The Navy in January 2015 estimated the average procurement cost of boats 2 through 12 in the Ohio replacement program at about $5.2 billion each in FY2010 dollars, and is working to reduce that figure to a

* This is an edited, reformatted and augmented version of a Congressional Research Service publication, No. R41129, dated March 24, 2015.

target of $4.9 billion each in FY2010 dollars. Even with this cost-reduction effort, observers are concerned about the impact the Ohio replacement program will have on the Navy's ability to procure other types of ships at desired rates in the 2020s and early 2030s.

Potential oversight issues for Congress for the Ohio replacement program include the following:

- the likelihood that the Navy will be able to reduce the average procurement cost of boats 2 through 12 in the program to the target figure of $4.9 billion each in FY2010 dollars;
- the accuracy of the Navy's estimate of the procurement cost of each SSBN(X);
- the prospective affordability of the Ohio replacement program and its potential impact on funding available for other Navy shipbuilding programs; and
- the question of which shipyard or shipyards will build SSBN(X)s.

This report focuses on the Ohio replacement program as a Navy shipbuilding program.

INTRODUCTION

This report provides background information and potential oversight issues for Congress on the Ohio replacement program (ORP), a program to design and build a new class of 12 ballistic missile submarines (SSBNs) to replace the Navy's current force of 14 Ohio-class SSBNs. The Ohio replacement program is also known as the SSBN(X) program. The Navy wants to procure the first Ohio replacement boat in FY2021, with advance procurement (AP) funding starting in FY2017. The Navy has identified the Ohio replacement program as its top priority program.

The Navy's proposed FY2015 budget requests $1,390.7 million for continued research and development work on the Ohio replacement program. Decisions that Congress makes on the Ohio replacement program could substantially affect U.S. military capabilities and funding requirements, and the U.S. shipbuilding industrial base.

This report focuses on the Ohio replacement program as a Navy shipbuilding program. Another CRS report discusses the SSBN(X) as an element of future U.S. strategic nuclear forces in the context of strategic nuclear arms control agreements.[1]

Background

U.S. Navy SSBNs in General

Mission of SSBNs

The U.S. Navy operates three kinds of submarines—nuclear-powered attack submarines (SSNs), nuclear-powered cruise missile submarines (SSGNs), and nuclear-powered ballistic missile submarines (SSBNs).[2] The SSNs and SSGNs are multi-mission ships that perform a variety of peacetime and wartime missions.[3] They do not carry nuclear weapons.[4]

The SSBNs, in contrast, perform a specialized mission of strategic nuclear deterrence. To perform this mission, SSBNs are armed with submarine-launched ballistic missiles (SLBMs), which are large, long-range missiles armed with multiple nuclear warheads. SSBNs launch their SLBMs from large-diameter vertical launch tubes located in the middle section of the boat.[5] The SSBNs' basic mission is to remain hidden at sea with their SLBMs, so as to deter a nuclear attack on the United States by another country by demonstrating to other countries that the United States has an assured second-strike capability, meaning a survivable system for carrying out a retaliatory nuclear attack.

Navy SSBNs, which are sometimes referred to informally as "boomers,"[6] form one leg of the U.S. strategic nuclear deterrent force, or "triad," which also includes land-based intercontinental ballistic missiles (ICBMs) and land-based long-range bombers. At any given moment, some of the Navy's SSBNs are conducting nuclear deterrent patrols. The Navy's report on its FY2011 30-year shipbuilding plan states: "These ships are the most survivable leg of the Nation's strategic arsenal and provide the Nation's only day-to-day assured nuclear response capability."[7] The Department of Defense's (DOD's) report on the 2010 Nuclear Posture Review (NPR), released on April 6, 2010, states that "strategic nuclear submarines (SSBNs) and the SLBMs they carry represent the most survivable leg of the U.S. nuclear Triad."[8]

Current Ohio-Class SSBNs

The Navy currently operates 14 Ohio (SSBN-726) class SSBNs. The boats are commonly called Trident SSBNs or simply Tridents because they carry Trident SLBMs.

A total of 18 Ohio-class SSBNs were procured in FY1974-FY1991. The ships entered service in 1981-1997. The boats were designed and built by

General Dynamics' Electric Boat Division (GD/EB) of Groton, CT, and Quonset Point, RI. They were originally designed for 30-year service lives but were later certified for 42-year service lives, consisting of two approximately 19- year periods of operation separated by an approximately four-year mid-life nuclear refueling overhaul, called an engineered refueling overhaul (ERO). The nuclear refueling overhaul includes both a nuclear refueling and overhaul work on the ship that is not related to the nuclear refueling.

Ohio-class SSBNs are designed to each carry 24 SLBMs, although by 2018, four SLBM launch tubes on each boat are to be deactivated, and the number of SLBMs that can be carried by each boat consequently is to be reduced to 20, so that the number of operational launchers and warheads in the U.S. force will comply with strategic nuclear arms control limits.

The first eight boats in the class were originally armed with Trident I C-4 SLBMs; the final 10 were armed with larger and more-capable Trident II D-5 SLBMs. The Clinton Administration's 1994 Nuclear Posture Review (NPR) recommended a strategic nuclear force for the START II strategic nuclear arms reduction treaty that included 14 Ohio-class SSBNs, all armed with D-5s. This recommendation prompted interest in the idea of converting the first four Ohio-class boats (SSBNs 726-729) into SSGNs, so as to make good use of the 20 years of potential operational life remaining in these four boats, and to bolster the U.S. SSN fleet. The first four Ohio-class boats were converted into SSGNs in 2002-2008,[9] and the next four (SSBNs 730-733) were backfitted with D-5 SLBMs in 2000-2005, producing the current force of 14 Ohio-class SSBNs, all of which are armed with D-5 SLBMs.

Eight of the 14 Ohio-class SSBNs are homeported at Bangor, WA, in Puget Sound; the other six are homeported at Kings Bay, GA, close to the Florida border.

Unlike most Navy ships, which are operated by single crews, Navy SSBNs are operated by alternating crews (called the Blue and Gold crews) so as to maximize the percentage of time that they spend at sea in deployed status. The Navy consequently maintains 28 crews to operate its 14 Ohio-class SSBNs.

The first of the 14 Ohio-class SSBNs (SSBN-730) will reach the end of its 42-year service life in 2027. The remaining 13 will reach the ends of their service lives at a rate of roughly one ship per year thereafter, with the 14th reaching the end of its service life in 2040.

The Navy has initiated a program to refurbish and extend the service lives of D-5 SLBMs to 2042 "to match the OHIO Class submarine service life."[10]

Figure 1 shows an Ohio-class SSBN with the hatches to some of its SLBM launch tubes open.

Source: U.S. Navy file photo accessed by CRS on February 24, 2011, at http://www.navy.mil/management photodb/photos/101029-N-1325N-005.jpg.

Figure 1. Ohio (SSBN-726) Class SSBN; With the hatches to some of its SLBM launch tubes open.

Summary of U.S. SSBN Designs

The Navy has operated four classes of SSBNs since 1959. **Table 1** compares the current Ohio-class SSBN design to the three earlier U.S. SSBN designs. As shown in the table, the size of U.S. SSBNs has grown over time, reflecting in part a growth in the size and number of SLBMs carried on each boat. The Ohio class carries an SLBM (the D-5) that is much larger than the SLBMs carried by earlier U.S. SSBNs, and it carries 24 SLBMs, compared to the 16 on earlier U.S. SSBNs.[11] In part for these reasons, the Ohio-class design, with a submerged displacement of 18,750 tons, is more than twice the size of earlier U.S. SSBNs.

U.S.-UK Cooperation on SLBMs and the New UK SSBN

SSBNs are also operated by the United Kingdom, France, Russia, China, and India. The UK's four Vanguard-class SSBNs, which entered service in 1993-1999, each carry 16 Trident II D-5 SLBMs. Previous classes of UK SSBNs similarly carried earlier-generation U.S. SLBMs.[12]

Table 1. U.S. SSBN Classes

	George Washington (SSBN-598) class	Ethan Allen (SSBN-608) class	Lafayette/Benjamin Franklin (SSBN-616/640) class	Ohio (SSBN-726) class
Number in class	5	5	31	18/14
Fiscal years procured	FY1958-FY1959	FY1959 and FY1961	FY1961-FY1964	FY1974/FY1977 -FY1991
Years in commission	1959-1985	1961-1992	1963-2002	1981/1984 - present
Length	381.7 feet	410.5 feet	425 feet	560 feet
Beam	33 feet	33 feet	33 feet	42 feet
Submerged displacement	6,700 tons	7,900 tons	8,250 tons	18,750 tons
Number of SLBM launch tubes	16	16	16	24 (to be reduced to 20 by 2018)
Final type(s) of SLBM carried	Polaris A-3	Polaris A-3	Poseidon C-3/	Trident II D-5
Diameter of those SLBMs	54 inches	54 inches	74 inches	83 inches
Length of those SLBMs	32.3 feet	32.3 feet	34 feet	44 feet
Weight of each SLBM (pounds)	36,000 pounds	36,000 pounds	65,000/73,000 pounds	~130,000 pounds
Range of SLBMs	~2,500 nm	~2,500 nm	~2,500 nm/~4,000 nm	~4,000 nm

Sources: Prepared by CRS based on data in Norman Polmar, *The Ships and Aircraft of the U.S. Fleet*, Annapolis, Naval Institute Press, various editions, and (for SSBN decommissioning dates) U.S. Naval Vessel Register.

Notes: Beam is the maximum width of a ship. For the submarines here, which have cylindrical hulls, beam is the diameter of the hull.

The range of an SLBM can vary, depending on the number and weight of nuclear warheads it carries; actual ranges can be lesser or greater than those shown.

The George Washington-class boats were procured as modifications of SSNs that were already under construction. Three of the boats were converted into SSNs toward the ends of their lives and were decommissioned in 1983-1985. The two boats that remained SSBNs throughout their lives were decommissioned in 1981.

All five Ethan Allen-class boats were converted into SSNs toward the ends of their lives. The boats were decommissioned in 1983 (two boats), 1985, 1991, and 1992.

Two of the Lafayette/Benjamin Franklin-class boats were converted into SSNs toward the ends of their lives and were decommissioned in 1999 and 2002. The 29 that remained SSBNs throughout their lives were decommissioned in 1986-1995. For 19 of the boats, the Poseidon C-3 was the final type of SLBM carried; for the other 12, the Trident I C-4 SLBM was the final type of SLBM carried.

A total of 18 Ohio-class SSBNs were built. The first four, which entered service in 1981-1984, were converted into SSGNs in 2002-2008. The remaining 14 boats entered service in 1984-1997. Although Ohio-class SSBNs are designed to each carry 24 SLBMs, by 2018, four SLBM launch tubes on each boat are to be deactivated, and the number of SLBMs that can be carried by each boat consequently is to be reduced to 20, so that the number of operational launchers and warheads in the U.S. force will comply with strategic nuclear arms control limits.

The UK's use of U.S.-made SLBMs on its SSBNs is one element of a long-standing close cooperation between the two countries on nuclear-related issues that is carried out under the 1958 Agreement for Cooperation on the Uses of Atomic Energy for Mutual Defense Purposes (also known as the Mutual Defense Agreement). Within the framework established by the 1958 agreement, cooperation on SLBMs in particular is carried out under the 1963 Polaris Sales Agreement and a 1982 Exchange of Letters between the two governments.[13] The Navy testified in March 2010 that "the United States and the United Kingdom have maintained a shared commitment to nuclear deterrence through the Polaris Sales Agreement since April 1963. The U.S. will continue to maintain its strong strategic relationship with the UK for our respective follow-on platforms, based upon the Polaris Sales Agreement."[14]

The first Vanguard-class SSBN was originally projected to reach the end of its service life in 2024, but an October 2010 UK defense and security review report states that the lives of the Vanguard class ships will now be extended by a few years, so that the four boats will remain in service into the late 2020s and early 2030s.[15]

The UK plans to replace the four Vanguard-class boats with three or four next-generation SSBNs called Successor class SSBNs. The October 2010 UK defense and security review report states that each new Successor class SSBN is to be equipped with 8 D-5 SLBMs, rather than 12 as previously planned. The report states that "'Initial Gate'—a decision to move ahead with early stages of the work involved—will be approved and the next phase of the project will start by the end of [2010]. 'Main Gate'—the decision to start building the submarines—is required around 2016."[16] The first new boat is to be delivered by 2028, or about four years later than previously planned.[17]

The UK has wanted the Successor SSBNs to carry D-5 SLBMs, and for any successor to the D-5 SLBM to be compatible with, or be capable of being made compatible with, the D-5 launch system. President George W. Bush, in a December 2006 letter to UK Prime Minister Tony Blair, invited the UK to participate in any program to replace the D-5 SLBMs, and stated that any successor to the D-5 system should be compatible with, or be capable of being made compatible with, the launch system for the D-5 SLBM.

The United States is assisting the UK with certain aspects of the Successor SSBN program. In addition to the modular Common Missile Compartment (CMC) discussed below (see "Common Missile Compartment (CMC)" in the following section on the Ohio replacement program), the United States is assisting the UK with the new PWR-3 reactor plant[18] to be used by the Successor SSBN. A December 2011 press report states that "there has been strong [UK] collaboration with the US [on the Successor program], particularly with regard to the CMC, the PWR, and other propulsion technology," and that the design concept selected for the Successor class employs "a new propulsion plant based on a US design, but using next-generation UK reactor technology (PWR-3) and modern secondary propulsion systems."[19] The U.S. Navy states that

> Naval Reactors, a joint Department of Energy/Department of Navy organization responsible for all aspects of naval nuclear propulsion, has an ongoing technical exchange with the UK Ministry of Defence under the US/UK 1958 Mutual Defence Agreement. The US/UK 1958 Mutual Defence Agreement is a Government to Government Atomic Energy Act agreement that allows the exchange of naval nuclear propulsion technology between the US and UK.
>
> Under this agreement, Naval Reactors is providing the UK Ministry of Defence with US naval nuclear propulsion technology to facilitate development of the naval nuclear propulsion plant for the UK's next generation SUCCESSOR ballistic missile submarine. The technology exchange is managed and led by the US and UK Governments, with participation from Naval Reactors prime contractors, private nuclear capable shipbuilders, and several suppliers. A UK based office comprised of about 40 US personnel provide full-time engineering support for the exchange, with additional support from key US suppliers and other US based program personnel as needed.
>
> The relationship between the US and UK under the 1958 mutual defence agreement is an ongoing relationship and the level of support varies depending on the nature of the support being provided. Naval Reactors work supporting the SUCCESSOR submarine is reimbursed by the UK Ministry of Defence.[20]

U.S. assistance to the UK on naval nuclear propulsion technology first occurred many years ago: To help jumpstart the UK's nuclear-powered submarine program, the United States transferred to the UK a complete nuclear propulsion plant (plus technical data, spares, and training) of the kind installed on the U.S. Navy's six Skipjack (SSN-585) class nuclear-powered attack submarines (SSNs), which entered service between 1959 and 1961. The plant was installed on the UK Navy's first nuclear-powered ship, the attack submarine *Dreadnought*, which entered service in 1963.

The December 2011 press report states that "the UK is also looking at other areas of cooperation between Successor and the Ohio Replacement Programme. For example, a collaboration agreement has been signed off regarding the platform integration of sonar arrays with the respective combat systems."[21]

Ohio Replacement Program

Program Origin and Early Milestones

Although the eventual need to replace the Ohio-class SSBNs has been known for many years, the Ohio replacement program can be traced more specifically to an exchange of letters in December 2006 between President George W. Bush and UK Prime Minister Tony Blair concerning the UK's desire to participate in a program to extend the service life of the Trident II D-5 SLBM into the 2040s, and to have its next-generation SSBNs carry D-5s. Following this exchange of letters, and with an awareness of the projected retirement dates of the Ohio-class SSBNs and the time that would likely be needed to develop and field a replacement for them, DOD in 2007 began studies on a next-generation sea-based strategic deterrent (SBSD).[22] The studies used the term sea-based strategic deterrent (SBSD) to signal the possibility that the new system would not necessarily be a submarine.

An Initial Capabilities Document (ICD) for a new SBSD was developed in early 2008[23] and approved by DOD's Joint Requirements Oversight Committee (JROC) on June 20, 2008.[24] In July 2008, DOD issued a Concept Decision providing guidance for an analysis of alternatives (AOA) for the program; an acquisition decision memorandum from John Young, DOD's acquisition executive, stated the new system would, barring some discovery, be a submarine.[25] The Navy established an Ohio replacement program office at about this same time.[26]

The AOA reportedly began in the summer or fall of 2008.[27] The AOA was completed, with final brief to the Office of the Secretary of Defense (OSD), on May 20, 2009. The final AOA report was completed in September 2009. An AOA Sufficiency Review Letter was signed by OSD's Director, Cost Assessment & Program Evaluation (CAPE) on December 8, 2009.[28] The AOA concluded that a new-design SSBN was the best option for replacing the Ohio-class SSBNs. (For a June 26, 2013, Navy blog post discussing options that were examined for replacing the Ohio-class SSBNs, see **Appendix A**.)

The program's Milestone A review meeting was held on December 9, 2010. On February 3, 2011, the Navy provided the following statement to CRS concerning the outcome of the December 9 meeting:

The OHIO Replacement Program achieved Milestone A and has been approved to enter the Technology Development Phase of the Dept. of Defense Life Cycle Management System as of Jan. 10, 2011.

This milestone comes following the endorsement of the Defense Acquisition Board (DAB), chaired by Dr. Carter (USD for Acquisition, Technology, and Logistics) who has signed the program's Milestone A Acquisition Decision Memorandum (ADM).

The DAB endorsed replacing the current 14 Ohio-class Ballistic Missile Submarines (SSBNs) as they reach the end of their service life with 12 Ohio Replacement Submarines, each comprising 16, 87-inch diameter missile tubes utilizing TRIDENT II D5 Life Extended missiles (initial loadout). The decision came after the program was presented to the Defense Acquisition Board (DAB) on Dec. 9, 2010.

The ADM validates the program's Technology Development Strategy and allows entry into the Technology Development Phase during which warfighting requirements will be refined to meet operational and affordability goals. Design, prototyping, and technology development efforts will continue to ensure sufficient technological maturity for lead ship procurement in 2019.[29]

Planned Procurement Quantity: 12 SSBN(X)s to Replace 14 Ohio-Class Boats

Navy plans call for procuring 12 SSBN(X)s to replace the current force of 14 Ohio-class SSBNs. In explaining the planned procurement quantity of 12 boats, the Navy states that 10 operational SSBNs—meaning boats not encumbered by lengthy maintenance actions—are needed to meet strategic nuclear deterrence requirements for having a certain number of SSBNs at sea at any given moment. The Navy states that a force of 14 Ohio-class boats was needed to meet this requirement because, during the middle years of the Ohio class life cycle, three and sometimes four of the boats are non-operational at any given moment on account of being in the midst of lengthy mid-life nuclear refueling overhauls or other extended maintenance actions. The Navy states that 12 rather than 14 SSBN(X)s will be needed to meet the requirement for 10 operational boats because the mid-life overhauls of SSBN(X)s, which will not include a nuclear refueling, will require less time (about two years) than the mid-life refueling overhauls of Ohio-class boats (which require about four years from contract award to delivery),[30] the result being that only two SSBN(X)s (rather than three or sometimes four) will be in the midst of mid-life overhauls or other extended maintenance actions at any given moment during the middle years of the SSBN(X) class life cycle.[31]

Procurement and Replacement Schedule

Table 2 shows the Navy's proposed schedule for procuring 12 SSBN(X)s, and for having SSBN(X)s replace Ohio-class SSBNs. As shown in **Table 2**, under the Navy's FY2012 budget, the first Ohio replacement boat was scheduled to be procured in FY2019, and Ohio replacement boats were to enter service on a schedule that would maintain the Navy's SSBN force at 12 boats.

As also shown in **Table 2**, the Navy's FY2013 budget deferred the procurement of the first Ohio replacement boat by two years, to FY2021. As a result of the deferment of the procurement of the lead boat from FY2019 to FY2021, the Navy's SSBN force will drop to 11 or 10 boats for the period FY2029-FY2041. The Navy states that the reduction to 11 or 10 boats during this period is acceptable in terms of meeting strategic nuclear deterrence requirements, because during these years, all 11 or 10 of the SSBNs in service will be operational (i.e., none of them will be in the midst of a lengthy mid-life overhaul). The Navy acknowledges that there is some risk in having the SSBN force drop to 11 or 10 boats, because it provides little margin for absorbing an unforeseen event that might force an SSBN into an unscheduled and lengthy maintenance action.[32] (See also the discussion above in "Planned Procurement Quantity: 12 SSBN(X)s to Replace 14 Ohio-Class Boats.")

The minimum level of 10 boats shown in **Table 2** for the period FY2032-FY2040 can be increased to 11 boats (providing some margin for absorbing an unforeseen event that might force an SSBN into an unscheduled and lengthy maintenance action) by accelerating by about one year the planned procurement dates of boats 2 through 12 in the program. Under this option, the second boat in the program would be procured in FY2023 rather than FY2024, the third boat in the program would be procured in FY2025 rather than FY2026, and so on. Implementing this option could affect the Navy's plan for funding the procurement of Virginia-class attack submarines during the period FY2022-FY2025.[33]

Table 2. Navy Schedule for Procuring SSBN(X)s and Replacing Ohio-Class SSBNs

	Schedule in FY2012 Budget				Schedule Under Subsequent Budgets			
Fiscal Year	Number of SSBN(X)s procured each year	Cumulative number of SSBN(X)s in service	Ohioclass SSBNs in service	Combined number of Ohio-class SSBNs and SSBN(X)s in service	Number of SSBN(X)s procured each year	Cumulative number of SSBN(X)s in service	Ohioclass SSBNs in service	Combined number of Ohio-class SSBNs and SSBN(X)s in service
2019	1		14	14			14	14
2020			14	14			14	14
2021			14	14	1		14	14
2022	1		14	14			14	14
2023			14	14			14	14
2024	1		14	14	1		14	14
2025	1		14	14			14	14
2026	1		14	14	1		14	14
2027	1		13	13	1		13	13
2028	1		12	13	1		12	12
2029	1	1	11	12	1		11	11
2030	1	2	10	12	1	1	10	11
2031	1	3	9	12	1	2	9	11
2032	1	4	8	12	1	2	8	10
2033	1	5	7	12	1	3	7	10
2034		6	6	12	1	4	6	10
2035		7	5	12	1	5	5	10
2036		8	4	12		6	4	10
2037		9	3	12		7	3	10
2038		10	2	12		8	2	10
2039		11	1	12		9	1	10
2040		12		12		10	0	10
2041		12		12		11	0	11
2042		12		12		12	0	12

Source: Navy FY2012-FY2015 budget submissions.

SSBN(X) Design Features

The design of the SSBN(X), now being developed, will reflect the following:

- The SSBN(X) is to be designed for a 42-year expected service life.[34]
- Unlike the Ohio-class design, which requires a mid-life nuclear refueling,[35] the SSBN(X) is to be equipped with a life-of-the-ship nuclear fuel core (a nuclear fuel core that is sufficient to power the ship for its entire expected service life).[36] Although the SSBN(X) will not need a mid-life nuclear refueling, it will still need a mid-life non-refueling overhaul (i.e., an overhaul that does not include a nuclear refueling) to operate over its full 40-year life.
- The SSBN(X) is to be equipped with an electric-drive propulsion train, as opposed to the mechanical-drive propulsion train used on other Navy submarines. The electric-drive system is expected to be quieter (i.e., stealthier) than a mechanical-drive system.[37]
- The SSBN(X) is to have SLBM launch tubes that are the same size as those on the Ohio class (i.e., tubes with a diameter of 87 inches and a length sufficient to accommodate a D-5 SLBM).
- The SSBN(X) will have a beam (i.e., diameter)[38] of 43 feet, compared to 42 feet on the Ohio-class design,[39] and a length of 560 feet, the same as that of the Ohio-class design.[40]
- Instead of 24 SLBM launch tubes, as on the Ohio-class design, the SSBN(X) is to have 16 SLBM launch tubes. (For further discussion of the decision to equip the boat with 16 tubes rather than 20, see **Appendix B**.)
- Although the SSBN(X) is to have fewer launch tubes than the Ohio-class SSBN, it is to be larger than the Ohio-class SSBN design, with a reported submerged displacement of 20,815 tons (as of August 2014), compared to 18,750 tons for the Ohio-class design.[41]
- The Navy states that "owing to the unique demands of strategic relevance, [SSBN(X)s] must be fitted with the most up-to-date capabilities and stealth to ensure they are survivable throughout their full 40-year life span."[42]

In an article published in June 2012, the program manager for the Ohio replacement program stated that "the current configuration of the Ohio replacement is an SSBN with 16 87-inchdiameter missile tubes, a 43-foot-diamater hull, electric-drive propulsion, [an] X-stern,[43] accommodations for

155 personnel, and a common submarine radio room[44] tailored to the SSBN mission."[45]

Acquisition Cost

A March 2015 GAO report assessing selected major DOD weapon acquisition programs states that the estimated total acquisition cost of the SSBN(X) program is $95,775.7 million (about $95.8 billion) in constant FY2015 dollars, including $11,801 million (about $11.8 billion) in research and development costs and $83,974.7 million (about $84.0 billion) in procurement costs.[46]

The Navy as of February 2015 estimates the procurement cost of the lead boat in the program at $14.5 billion in then-year dollars, including $5.7 billion in detailed design and nonrecurring engineering (DD/NRE) costs for the entire class, and $8.8 billion in construction costs for the ship itself. (It is a traditional budgeting practice for Navy shipbuilding programs to attach the DD/NRE costs for a new class of ships to the procurement cost of the lead ship in the class.) In constant FY2010 dollars, these figures become $10.4 billion, including $4.2 billion in DD/NRE costs and $6.2 billion in construction costs for the ship itself.[47]

The Navy in February 2010 preliminarily estimated the procurement cost of each Ohio replacement boat at $6 billion to $7 billion in FY2010 dollars.[48] Following the Ohio replacement program's December 9, 2010, Milestone A acquisition review meeting (see "Program Origin and Early Milestones"), DOD issued an Acquisition Decision Memorandum (ADM) that, among other things, established a target average unit procurement cost for boats 2 through 12 in the program of $4.9 billion in constant FY2010 dollars.[49] The Navy is working to achieve this target cost. In January 2015, the Navy stated that its cost-reduction efforts had reduced the estimated average unit procurement cost of boats 2 through 12 to about $5.2 billion each in constant FY2010 dollars.[50] The Navy continues examining potential further measures to bring the cost of boats 2 through 12 closer to the $4.9 billion target cost.

The above cost figures do not include costs for refurbishing D-5 SLBMs so as to extend their service lives to 2042.

Operation and Support (O&S) Cost

The Navy is working to reduce the estimated operation and support (O&S) cost of each SSBN(X) from $124 million per year to $110 million per year in constant FY2010 dollars.[51]

Common Missile Compartment (CMC)

Current U.S. and UK plans call for the SSBN(X) and the UK's Successor SSBN to use a missile compartment—the middle section of the boat with the SLBM launch tubes—of the same general design.[52] As mentioned earlier (see "U.S.-UK Cooperation on SLBMs"), the UK's SSBN is to be armed with eight SLBMs, or half the number to be carried by the SSBN(X). The modular design of the CMC will accommodate this difference. Since the UK's first Vanguard-class SSBN was originally projected to reach the end of its service life in 2024—three years before the first Ohio-class SSBN is projected to reach the end of its service life—design work on the CMC began about three years sooner than would have been required to support the Ohio replacement program alone. This is the principal reason why the FY2010 budget included a substantial amount of research and development funding for the CMC. The UK is providing some of the funding for the design of the CMC, including a large portion of the initial funding.

A March 2010 Government Accountability office (GAO) report stated:

> According to the Navy, in February 2008, the United States and United Kingdom began a joint effort to design a common missile compartment. This effort includes the participation of government officials from both countries, as well as industry officials from Electric Boat Corporation and BAE Systems. To date, the United Kingdom has provided a larger share of funding for this effort, totaling just over $200 million in fiscal years 2008 and 2009.[53]

A March 2011 GAO report stated:

> The main focus of OR [Ohio Replacement program] research and development to date has been the CMC. The United Kingdom has provided $329 million for this effort since fiscal year 2008. During fiscal years 2009 and 2010, the Navy had allocated about $183 million for the design and prototyping of the missile compartment.[54]

A May 2010 press report stated that "the UK has, to date, funded the vast majority of [the CMC's] upfront engineering design activity and has established a significant presence in Electric Boat's Shaw's Cove CMC design office in New London, CT."[55]

Under the October 2010 UK defense and security review report (see "U.S.-UK Cooperation on SLBMs"), the UK now plans to deliver its first Successor class SSBN in 2028, or about four years later than previously planned.

Program Funding

Table 3 shows funding for the Ohio replacement program. The table shows U.S. funding only; it does not include funding provided by the UK to help pay for the design of the CMC. As can be seen in the table, the Navy's proposed FY2016 budget requests $1,390.7 million for continued research and development work on the program.

Table 3. Ohio Replacement Program Funding (Millions of then-year dollars, rounded to nearest tenth; totals may not add due to rounding)

	FY15	**FY16 (req.)**	**FY17 (proj.)**	**FY18 (proj.**	**FY19 (proj.)**	**FY20 (proj.)**
Research and development (R&D) funding						
PE0603570N/Project 3219	370.0	419.3	408.1	398.7	289.0	276.7
PE0603595N/Project 3220	816.8	971.4	712.4	766.5	484.9	206.0
PE0603595N/Project 3237	36.5	0	0	0	0	0
Subtotal R&D funding	**1,223.3**	**1,390.7**	**1,120.5**	**1,165.2**	**773.9**	**482.7**
Military Construction (MilCon) funding	**24.3**	**0**	**0**	**0**	**0**	**0**
Procurement funding	**0**	**0**	**777.8**	**791.8**	**2,771.3**	**1,316.3**
TOTAL	**1,247.6**	**1,390.7**	**1,898.3**	**1,957.0**	**1,547.8**	**965.4**

Source: Navy FY2016 budget submission.

Notes: **PE** means Program Element, that is, a research and development line item. A Program Element may include several projects. **PE0603570N/Project 3219** is SSBN(X) reactor plant project within the PE for Advanced Nuclear Power Systems. **PE0603561N/Project 3220** is Sea-Based Strategic Deterrent (SBSD) Advanced Submarine System Development project within the PE for Ohio Replacement. **PE0603595N/Project 3237** is Launch Test Facility project within the PE for Ohio Replacement. **Military Construction (MilCon) funding** for FY2015 is for an Ohio replacement program launch test facility (MilCon/0805376N) ($23.985 million) and Ohio Replacement Power and Propulsion Facility (MilCon/0901211N, design funds) ($0.364 million). **Procurement funding** shown in FY2017 through FY2020 is advance procurement funding for first SSBN(X), which is scheduled to be procured in FY2021.

ISSUES FOR CONGRESS

Likelihood That Navy Will Reach $4.9 Billion Target Cost

One potential oversight issue for Congress regarding the Ohio replacement program is the likelihood that the Navy will be able to achieve

DOD's goal of reducing the average unit procurement cost of boats 2 through 12 in the program to $4.9 billion each in FY2010 dollars. As mentioned earlier, as of January 2015, the Navy estimated that its cost-reduction efforts had reduced the average unit procurement cost of boats 2 through 12 to about $5.2 billion each in FY2010 dollars, leaving another $300 million or so in cost reduction to reach the $4.9 billion target cost.

A January 26, 2015, press report quoted Rear Admiral David Johnson, the program executive officer for submarines, as stating that in achieving the targeted reduction in per-boat procurement cost, "I'm confident we'll get to the $4.9 billion number that we have [as a target], we just have to keep working at it and we'll need the help of Congress with multiyear authorities in how we'll actually fund the ships."[56]

Potential oversight questions include the following:

- How did DOD settle on the figure of $4.9 billion in FY2010 dollars as the target average unit procurement cost for boats 2 through 12 in the program? On what analysis was the selection of this figure based?
- How difficult will it be for the Navy to reach this target cost? What options is the Navy examining to achieve the additional $300 million or so in unit procurement cost savings needed to reach it?
- Would a boat costing $4.9 billion have sufficient capability to perform its intended missions?
- What, if anything, does DOD plan to do if the Navy is unable to achieve the $4.9 billion target cost figure? If $4.9 billion is the target figure, is there a corresponding "ceiling" figure higher than $4.9 billion, above which DOD would not permit the Ohio replacement program to proceed? If no such figure exists, should DOD establish one?

Accuracy of Navy's Estimated Unit Procurement Cost

Overview

Another potential oversight issue for Congress concerns the accuracy of the Navy's estimate of the procurement cost of each SSBN(X). The accuracy of the Navy's estimate is a key consideration in assessing the potential affordability of the Ohio replacement program, including its potential impact on the Navy's ability to procure other kinds of ships during the years of SSBN(X) procurement. Some of the Navy's ship designs in recent years, such as the Gerald R. Ford (CVN-78) class aircraft carrier,[57] the San Antonio (LPD-17) class amphibious ship[58] and the Littoral Combat Ship (LCS),[59] have proven to be substantially more expensive to build than the Navy originally estimated.

The accuracy of the Navy's estimate can be assessed in part by examining known procurement costs for other recent Navy submarines—including Virginia (SSN-774) class attack submarines (which are currently being procured), Seawolf (SSN-21) class attack submarines (which were procured prior to the Virginia class), and Ohio (SSBN-726) class ballistic missile submarines— and then adjusting these costs for the Ohio replacement program so as to account for factors such as differences in ship displacement and design features, changes over time in submarine technologies (which can either increase or reduce a ship's procurement cost, depending on the exact technologies in question), advances in design for producibility (i.e., design features that are intended to make ships easier to build), advances in shipyard production processes (such as modular construction), and changes in submarine production economies of scale (i.e., changes in the total number of attack submarines and ballistic missile submarines under construction at any one time).

The Navy's estimated unit procurement cost for the program at any given point will reflect assumptions on, among other things, which shipyard or shipyards will build the boats, and how much Virginia-class construction will be taking place in the years when SSBN(X)s are being built. Changing the Navy's assumption about which shipyard or shipyards will build SSBN(X)s could reduce or increase the Navy's estimated unit procurement cost for the boats. If shipbuilding affordability pressures result in Virginia-class boats being removed from the 30-year shipbuilding plan during the years of SSBN(X) procurement, the resulting reduction in submarine production economies of scale could make SSBN(X)s more expensive to build than the Navy estimates.

December 2014 CBO Report

A December 2014 Congressional Budget Office (CBO) report on the cost of the Navy's shipbuilding programs stated:

> The design, cost, and capabilities of the Ohio Replacement submarine class are among the most significant uncertainties in the Navy's and CBO's analyses of the cost of future shipbuilding....

The Navy currently estimates the cost of the first Ohio Replacement submarine at $12.4 billion in 2014 dollars. The estimated average cost of follow-on ships is now $6.0 billion, which implies a total cost for 12 submarines of $79 billion, or an average of $6.6 billion each. However, the Navy has stated an objective of reducing that $6.0 billion figure to $5.5 billion.

The Navy's estimate represents considerable improvement in the cost per thousand tons of the first Ohio Replacement submarine compared with the first Virginia submarine—an improvement that would affect costs for the entire new class of ballistic missile submarines. The main reason for those improved costs by weight for the Ohio Replacement is that the Navy will recycle, to the extent possible, the design, technology, and parts used for the Virginia class. Furthermore, because ballistic missile submarines (like the Ohio Replacements) tend to be larger and less dense ships than attack submarines (like the Virginia class), they will be easier to build and therefore less expensive on a weight basis, the Navy argues.

Yet the historical record for the lead ships of new classes of submarines in the 1970s and 1980s provides little reason to believe that ballistic missile submarines are cheaper by weight to build than attack submarines.... The first Ohio class submarine was more expensive to build than the lead ships of the two classes of attack submarines built during the same period—the Los Angeles and the Improved Los Angeles (which made design changes to the Los Angeles that included the addition of 12 vertical launch system cells). In addition, the average costs by weight of the first 12 or 13 ships of the Ohio, Los Angeles, and Improved Los Angeles classes were virtually identical. By the 1990s, the costs of lead ships for submarines had grown substantially, with both the first Seawolf submarine and first Virginia submarine costing about the same by weight—even though the Seawolf is 20 percent larger and was built nine years earlier than the first Virginia.

Using data from the Virginia class submarine program, CBO estimates that the first Ohio Replacement submarine will cost $13.8 billion in 2014 dollars. Estimating the cost of the first submarine of a class with an entirely new design is particularly difficult because of uncertainty about how much the Navy will spend on nonrecurring engineering and detail design. All told, 12 Ohio Replacement submarines would cost $92 billion in CBO's estimation, or an average of $7.7 billion each—$1.1 billion more per boat than the Navy's estimate. That average includes the $13.8 billion estimated cost of the lead submarine and a $7.1 billion average estimated cost for the 2nd through 12th submarines. Research and development would cost an additional $10 billion to $15 billion, for a total program cost of $102 billion to $107 billion, CBO estimates.

Overall, the Navy expects a 19 percent improvement in the cost-to-weight relationship of the Ohio Replacement class compared with the first 12 of the Virginia class. CBO estimates a 4 percent improvement, which is based in part on savings that will probably be achieved because of projected concurrent production of the Ohio Replacement and Virginia class submarines.[60]

Program Affordability and Impact on Other Navy Shipbuilding Programs

Overview

Another oversight issue for Congress concerns the prospective affordability of the Ohio replacement program and its potential impact on funding available for other Navy shipbuilding programs. It has been known for some time that the Ohio replacement program, if funded through the Navy's shipbuilding account, could make it considerably more difficult for the Navy to procure other kinds of ships in desired numbers, unless the shipbuilding account were increased to accommodate the additional funding needs of the Ohio replacement program. On February 26, 2015, Admiral Jonathan Greenert, the Chief of Naval Operations, testified that

> In the long term beyond 2020, I am increasingly concerned about our ability to fund the Ohio Replacement ballistic missile submarine (SSBN) program—our highest priority program— within our current and projected resources. The Navy cannot procure the Ohio Replacement in the 2020s within historical shipbuilding funding levels without severely impacting other Navy programs.[61]

On February 25, 2015, Department of the Navy officials testified that

> The Navy continues to need significant increases in our topline beyond the FYDP [Future Years Defense Plan], not unlike that during the period of [the original] Ohio [class] construction [effort], in order to afford the OR [Ohio replacement] SSBN procurement costs. Absent a significant increase to the SCN [Shipbuilding and Conversion, Navy] appropriation [i.e., the Navy's shipbuilding account], OR SSBN construction will seriously impair construction of virtually all other ships in the battle force: attack submarines, destroyers, and amphibious warfare ships. The shipbuilding industrial base will be commensurately impacted and shipbuilding costs would spiral unfavorably. The resulting battle force would fall markedly short of the FSA [Force Structure Assessment—the Navy's force structure goal for the fleet as a whole], [and be] unable to meet fleet inventory requirements. The National

> Sea-Based Deterrence Fund [see discussion below] is a good first step in that it acknowledges the significant challenge of resourcing the OR SSBN, but the fund is unresourced [i.e., no funding has been placed into the account].[62]

Ohio Replacement Program Is Navy's Top Priority Program

On September 18, 2013, Admiral Jonathan Greenert, the Chief of Naval Operations, testified that the Ohio replacement program "is the top priority program for the Navy."[63] Navy officials since then have reiterated this statement.

The Navy's decision to make the Ohio replacement program its top program priority means that the Ohio replacement program will be fully funded, and that any resulting pressures on the Navy's shipbuilding account would be borne by other Navy programs, including shipbuilding programs. At a September 12, 2013, hearing before the Seapower and Projection Forces subcommittee of the House Armed Services Committee on undersea warfare, a Navy official stated:

> The CNO has stated, his number one priority as the chief of Naval operations, is our—our strategic deterrent—our nuclear strategic deterrent. That will trump all other vitally important requirements within our Navy, but if there's only one thing that we do with our ship building account, we—we are committed to sustaining a two ocean national strategic deterrent that protects our homeland from nuclear attack, from other major war aggression and also access and extended deterrent for our allies.[64]

At this same hearing, Navy officials testified that the service is seeking about $4 billion per year over 15 years in supplemental funding—a total of about $60 billion—for the Ohio replacement program.[65] The 15 years in question, Navy officials suggested in their testimony, are the years in which the Ohio replacement boats are to be procured (FY2021-FY2035, as shown in **Table 2**).[66] The $60 billion in additional funding equates to an average of $5 billion for each of the 12 boats, which is close to the Navy's target of an average unit procurement cost of $4.9 billion in constant FY2010 dollars for boats 2 through 12 in the program. The Navy stated at the hearing that the $60 billion in supplemental funding that the Navy is seeking would equate to less than 1% of DOD's budget over the 15-year period. The Navy also suggested that the 41 pre-Ohio class SSBNs that were procured in the 1950s and 1960s (see **Table 1**) were partially financed with funding that was provided as a supplement to the Navy's budget.[67]

The Navy officials stated at the September 12 hearing that if the Navy were to receive about $30 billion in supplemental funding for the Ohio replacement program—about half the amount that the Navy is requesting—then the Navy would need to eliminate from its 30-year shipbuilding plan a notional total of 16 other ships, including, notionally, 4 Virginia-class attack submarines, 4 destroyers, and 8 other combatant ships (which might mean ships such as Littoral Combat Ships or amphibious ships). Navy officials stated, in response to a question, that if the Navy were to receive none of the supplemental funding that it is requesting, then these figures could be doubled—that is, that the Navy would need to eliminate from its 30-year shipbuilding plan a notional total of 32 other ships, including, notionally, 8 Virginia-class attack submarines, 8 destroyers, and 16 other combatant ships.[68]

National Sea-Based Deterrence Fund

As an apparent measure for mitigating the impact of the Ohio replacement program on the Navy's ability to fund other priorities (particularly other shipbuilding programs), Congress, as part of its markup of the Navy's proposed FY2015 budget, created a National Sea-Based Deterrence Fund that will be separate from the Navy's regular shipbuilding account. The fund was created by Section 1022 of the Carl Levin and Howard P. "Buck" McKeon National Defense Authorization Act for Fiscal Year 2015 (H.R. 3979/P.L. 113-291 of December 19, 2014), which states:

> SEC. 1022. NATIONAL SEA-BASED DETERRENCE FUND. (a) Establishment of Fund.—
>
> (1) In general.—Chapter 131 of title 10, United States Code, is amended by inserting after section 2218 the following new section:
>
> "Sec. 2218a. National Sea-Based Deterrence Fund
>
> (a) Establishment.—There is established in the Treasury of the United States a fund to be known as the 'National Sea-Based Deterrence Fund'.
>
> (b) Administration of Fund.—The Secretary of Defense shall administer the Fund consistent with the provisions of this section.
>
> (c) Fund Purposes.—(1) Funds in the Fund shall be available for obligation and expenditure only for construction (including design of vessels), purchase, alteration, and conversion of national sea-based deterrence vessels.
>
> "(2) Funds in the Fund may not be used for a purpose or program unless the purpose or program is authorized by law.
>
> (d) Deposits.—There shall be deposited in the Fund all funds appropriated to the Department of Defense for construction (including design

of vessels), purchase, alteration, and conversion of national sea-based deterrence vessels.

(e) Expiration of Funds After 5 Years.—No part of an appropriation that is deposited in the Fund pursuant to subsection (d) shall remain available for obligation more than five years after the end of fiscal year for which appropriated except to the extent specifically provided by law.

(f) Budget Requests.—Budget requests submitted to Congress for the Fund shall separately identify the amount requested for programs, projects, and activities for construction (including design of vessels), purchase, alteration, and conversion of national sea-based deterrence vessels.

(g) Definitions.—In this section:

(1) The term 'Fund' means the National Sea-Based Deterrence Fund established by subsection (a).

(2) The term 'national sea-based deterrence vessel' means any vessel owned, operated, or controlled by the Department of Defense that carries operational intercontinental ballistic missiles.".

(2) Clerical amendment.—The table of sections at the beginning of chapter 131 of such title is amended by inserting after the item relating to section 2218 the following new item:

"2218a. National Sea-Based Deterrence Fund.". (b) Transfer Authority.—

(1) In general.—Subject to paragraph (2), and to the extent provided in appropriations Acts, the Secretary of Defense may transfer to the National Sea-Based Deterrence Fund established by section 2218a of title 10, United States Code, as added by subsection (a)(1), amounts not to exceed $3,500,000,000 from unobligated funds authorized to be appropriated for fiscal years 2014, 2015, or 2016 for the Navy for the Ohio Replacement Program. The transfer authority provided under this paragraph is in addition to any other transfer authority provided to the Secretary of Defense by law.

(2) Availability.—Funds transferred to the National Sea-Based Deterrence Fund pursuant to paragraph (1) shall remain available for the same period for which the transferred funds were originally appropriated.

Prior to this legislation, some observers had suggested funding the procurement of SSBN(X)s outside the Navy's shipbuilding budget, so as to preserve Navy shipbuilding funds for other Navy shipbuilding programs. There was some precedent for such an arrangement:

- Construction of DOD sealift ships and Navy auxiliary ships has been funded in recent years in the National Defense Sealift Fund (NDSF), a part of DOD's budget that is outside the Shipbuilding and Conversion,

Navy (SCN) appropriation account, and also outside the procurement title of the DOD appropriations act.

- Most spending for ballistic missile defense (BMD) programs (including procurement-like activities) is funded through the Defense-Wide research and development account rather than through the research and development and procurement accounts of the individual military services.

A rationale for funding DOD sealift ships in the NDSF has been that DOD sealift ships perform a transportation mission that primarily benefits services other than the Navy, and therefore should not be forced to compete for funding in a Navy budget account that funds the procurement of ships central to the Navy's own missions. A rationale for funding BMD programs together in the Defense-Wide research and development account is that this makes potential tradeoffs in spending among various BMD programs more visible and thereby helps to optimize the use of BMD funding.

As a reference tool for better understanding DOD spending, DOD includes in its annual budget submission a presentation of the DOD budget reorganized into 11 program areas, of which one is strategic forces. The FY2015 budget submission, for example, shows that about $11.7 billion is requested for strategic forces for FY2015.[69]

Some Options for Further Addressing the Issue

In addition to creating the National Sea-Based Deterrent Fund and making further changes and refinements in the design of the SSBN(X), options for further reducing the cost of the Ohio replacement program and the program's potential impact on funding available for other Navy programs (particularly shipbuilding programs) include the following:

- using a joint block buy contract that would cover both the Ohio replacement program and the Virginia-class attack submarine program;
- altering the schedule for procuring the SSBN(X)s so as to create additional opportunities for using incremental funding for procuring the ships; and
- reducing the planned number of SSBN(X)s. Each of these options is discussed below.

Joint Block Buy Contract Covering Both Ohio Replacement and Virginia-Class Programs

To help reduce ship procurement costs, the Navy in recent years has made extensive use of multiyear procurement (MYP) contracts and block buy contracts in its shipbuilding programs,[70] including the Virginia class attack submarine program.[71] In light of this, the Navy will likely seek to use block buy and/or MYP contracting in the Ohio replacement program. Beyond that, the Navy is investigating the possibility of using a single, joint-class block buy contract that would cover both Ohio replacement boats and Virginia class boats. Such a contract, which could be viewed as precedent-setting in its scope, could offer savings beyond what would be possible using separate MYP or block buy contracts for the two submarine programs. A March 2014 GAO report stated that if the Navy decides to propose such a contract, it would develop a legislative proposal in 2017.[72] The Navy reportedly plans to finalize its acquisition strategy for the Ohio replacement program, including the issue of the contracting approach to be used, in the fall of 2016 as part of DOD's Milestone B decision for the program.[73]

Altering Procurement Schedule to Make More Use of Incremental Funding

Another option for managing the potential impact of the Ohio replacement program on other Navy shipbuilding programs would be to stretch out the schedule for procuring SSBN(X)s and make greater use of split funding (i.e., two-year incremental funding) in procuring them.[74] This option would not reduce the total procurement cost of the Ohio replacement program—to the contrary, it might increase the program's total procurement cost somewhat by reducing production learning curve benefits in the Ohio replacement program.[75] This option could, however, reduce the impact of the Ohio replacement program on the amount of funding available for the procurement of other Navy ships in certain individual years. This might reduce the amount of disruption that the Ohio replacement program causes to other shipbuilding programs in those years, which in turn might avoid certain disruption-induced cost increases for those other programs. The annual funding requirements for the Ohio replacement program might be further spread out by funding some of the SSBN(X)s with three- or four-year incremental funding.[76]

Table 4 shows the Navy's currently planned schedule for procuring 12 SSBN(X)s and a notional alternative schedule that would start two years earlier and end two years later than the Navy's currently planned schedule. Although the initial ship in the alternative schedule would be procured in

FY2019, it could be executed as it if were funded in FY2021. Subsequent ships in the alternative schedule that are funded earlier than they would be under the Navy's currently planned schedule could also be executed as if they were funded in the year called for under the Navy's schedule. Congress in the past has funded the procurement of ships whose construction was executed as if they had been procured in later fiscal years.[77] The ability to stretch the end of the procurement schedule by two years, to FY2035, could depend on the Navy's ability to carefully husband the use of the nuclear fuel cores on the last two Ohio-class SSBNs, so as to extend the service lives of these two ships by one or two years. Alternatively, Congress could grant the Navy the authority to begin construction on the 11th boat a year before its nominal year of procurement, and the 12th boat two years prior to its nominal year of procurement.

Table 4. Navy SSBN(X) Procurement Schedule and a Notional Alternative Schedule

Fiscal year	Navy's Schedule	Boat might be particularly suitable for 2-, 3-, or 4-year incremental funding	Notional alternative schedule	Boat might be particularly suitable for 2-, 3-, or 4-year incremental funding
2019			1	X
2020				
2021	1	X	1	X
2022				
2023			1	X
2024	1	X		
2025			1	X
2026	1			
2027	1		1	
2028	1		1	
2029	1		1	
2030	1		1	
2031	1		1	X
2032	1			
2033	1	X	1	X
2034	1	X		
2035	1	X	1	X
2036				
2037			1	X
Total	**12**		**12**	

Source: Navy's current plan is taken from the Navy's FY2015 budget submission. Potential alternative plan prepared by CRS.

Notes: Notional alternative schedule could depend on Navy's ability to carefully husband the use of the nuclear fuel cores on the last two Ohio-class SSBNs, so as to extend the service lives of these two ships by one or two years. Alternatively, Congress could grant the Navy the authority to begin construction on the 11th boat a year before its nominal year of procurement, and the 12th boat two years prior to its nominal year of procurement. Under Navy's schedule, the boat to be procured in FY2033 might be particularly suitable for 4-year incremental funding, and boat to be procured in FY2034 might be particularly suitable for 3- or 4-year incremental funding.

A December 19, 2011, press report states:

> The Office of Management and Budget's Nov. 29[, 2011,] passback memorandum to the Defense Department [regarding the FY2013 DOD budget] warns that the effort to build replacements for aging Ohio-class submarines is not exempt from rules requiring each new vessel to be fully funded in a single year....
>
> Spreading the cost of a big-ticket ship over more than one year—an approach referred to as "incremental funding"—is only allowed when a program meets three criteria, OMB writes....
>
> "OMB does not anticipate that the OHIO Replacement program will meet these criteria," the passback memo states.[78]

Reducing the Planned Number of SSBN(X)s

Some observers over the years have advocated or presented options for an SSBN force of fewer than 12 SSBNs. A November 2013 CBO report on options for reducing the federal budget deficit, for example, presented an option for reducing the SSBN force to eight boats as a cost-reduction measure.[79] Earlier CBO reports have presented options for reducing the SSBN force to 10 boats as a cost-reduction measure.[80] CBO reports that present such options also provide notional arguments for and against the options. A June 2010 report by a group known as the Sustainable Defense Task Force recommends reducing the SSBN force to 7 boats;[81] a September 2010 report from the Cato Institute recommends reducing the SSBN force to 6 boats,[82] and a September 2013 report from a group organized by the Stimson Center recommends reducing the force to 10 boats.[83]

Views on whether a force of fewer than 12 SSBN(X)s would be adequate could depend on, among other things, assessments of strategic nuclear threats to the United States and the role of SSBNs in deterring such threats as a part of overall U.S. strategic nuclear forces, as influenced by the terms of strategic nuclear arms control agreements.[84] Reducing the number of SSBNs below 12 could also raise a question as to whether the force should continue to be homeported at both Bangor, WA, and Kings Bay, GA, or consolidated at a single location.

U.S. strategic nuclear deterrence plans require a certain number of strategic nuclear warheads to be available for use on a day-to-day basis. After taking into account warheads on the other two legs of the strategic nuclear triad, the number of warheads on an SSBN's SLBMs, and factors independent of the number of warheads on the SLBMs, this translates into a requirement for a certain number of SSBNs to be on station (i.e., within range of expected targets) in Pacific and Atlantic waters at any given moment. The SSBN force is sized to support this requirement. Given the time needed for at-sea training operations, restocking SSBNs with food and other consumables, performing maintenance and repair work on the SSBNs, and transiting to and from deterrent patrol areas, only a fraction of the SSBN force can be on patrol at any given moment. The Navy's position (see "Planned Procurement Quantity: 12 SSBN(X)s to Replace 14 Ohio-Class Boats" in "Background") is that the requirement for having a certain number of SSBNs on patrol at any given moment translates into a need for a force of 14 Ohio-class boats, and that this requirement can be met in the future by a force of 12 SSBN(X)s.

Construction Shipyard(s)

Another potential issue for Congress regarding the Ohio replacement program is which shipyard or shipyards would build SSBN(X)s. Two U.S. shipyards are capable of building nuclear-powered submarines—General Dynamics' Electric Boat Division (GD/EB) of Groton, CT, and Quonset Point, RI, and Newport News Shipbuilding (NNS), of Newport News, VA, which forms part of Huntington Ingalls Industries (HII). GD/EB's primary business is building nuclear-powered submarines; it can also perform submarine overhaul work. NNS's primary lines of business are building nuclear-powered aircraft carriers, building nuclear-powered submarines, and performing overhaul work on nuclear-powered aircraft carriers. The Navy reportedly plans to finalize its acquisition strategy for the Ohio replacement program, including the issue of which shipyard or shipyards will build the boats, in the fall of 2016 as part of DOD's Milestone B decision for the program.[85]

Table 5. Construction Shipyards of U.S. SSBNs

	George Washington (SSBN-598) class	**Ethan Allen (SSBN-608) class**	**Lafayette/ Benjamin Franklin (SSBN-616/640) class**	**Ohio (SSBN-726) class**
Fiscal years procured	FY58-FY59	FY59 and FY61	FY61-FY64	FY77-FY91
Number built by GD/EB	2	2	13	18
Number built by NNS	1	3	10	
Number built by Mare Island NSY	1		6	
Number built by Portsmouth NSY	1		2	
Total number in class	**5**	**5**	**31**	**18**

Source: Prepared by CRS based on data in Norman Polmar, *The Ships and Aircraft of the U.S. Fleet*, Annapolis, Naval Institute Press, various editions. NSY means naval shipyard.

Notes: GD/EB was the builder of the first boat in all four SSBN classes. The George Washington-class boats were procured as modifications of SSNs that were already under construction. A total of 18 Ohio-class SSBNs were built; the first four were converted into SSGNs in 2002-2008, leaving 14 in service as SSBNs.

Table 5 shows the numbers of SSBNs built over time by GD/EB, NNS, and two government-operated naval shipyards (NSYs)—Mare Island NSY, located in the San Francisco Bay area, and Portsmouth NSY of Portsmouth, NH, and Kittery, ME. Mare Island NSY is no longer in operation. NSYs have not built new Navy ships since the early 1970s; since that time, they have focused solely on overhauling and repairing Navy ships.

As can be seen in the table, the Ohio-class boats were all built by GD/EB, and the three previous SSBN classes were built partly by GD/EB, and partly by NNS. GD/EB was the builder of the first boat in all four SSBN classes. The most recent SSBNs built by NNS were the *George C. Marshall* (SSBN-654) and *George Washington Carver* (SSBN-656), which were Lafayette/Benjamin Franklin-class boats that were procured in FY1964 and entered service in 1966.

There are at least five basic possibilities for building SSBN(X)s:

- **build all SSBN(X)s at GD/EB**—the approach that was used for building the Ohio-class SSBNs;

- **build all SSBN(X)s at NNS**;
- **build some SSBN(X)s GD/EB and some at NNS**—the approach that was used for building the George Washington-, Ethan Allen-, and Lafayette/Benjamin Franklin class SSBNs;
- **build each SSBN(X) jointly at GD/EB and NNS, with final assembly of the boats alternating between the yards**—the approach currently being used for building Virginia-class SSNs;[86] and
- **build each SSBN(X) jointly at GD/EB and NNS, with one yard—either GD/EB or NNS—performing final assembly on every boat**.

In assessing these five approaches, policymakers may consider a number of factors, including their potential costs, their potential impacts on employment levels at GD/EB and NNS, and the relative value of preserving SSBN-unique construction skills (such as those relating to the construction and installation of SLBM compartments) at one shipyard or two. The relative costs of these five approaches could depend on a number of factors, including the following:

- each yard's share of SSBN(X) production work (if both yards are involved);
- the number of SSNs procured during the years of SSBN(X) procurement (which can affect economies of scale in submarine production);
- whether the current joint-production arrangement for the Virginia class remains in effect during those years;[87] and
- the volume of non-submarine-construction work performed at the two shipyards during these years, which would include in particular aircraft carrier construction and overhaul work at NNS.

LEGISLATIVE ACTIVITY FOR FY2016

FY2016 Funding Request

As shown in **Table 3**, the Navy's proposed FY2016 budget requests $1,390.7 million in research and development funding for the Ohio replacement program, including $419.3 million for Project 3219 (SSBN[X] reactor plant) in PE0603570N (Advanced Nuclear Power Systems) and $971.4

million for Project 3220 (SBSD [Sea-Based Strategic Deterrent] Advanced Submarine System Development) in PE0603595N (Ohio Replacement).

APPENDIX A. JUNE 2013 NAVY BLOG POST REGARDING OHIO REPLACEMENT OPTIONS

This appendix presents the text of a June 26, 2013, blog post by Rear Admiral Richard Breckenridge, the Navy's Director for Undersea Warfare (N97), discussing options that were examined for replacing the Ohio-class SSBNs. The text is as follows:

> Over the last five years, the Navy – working with U.S. Strategic Command, the Joint Staff and the Office of the Secretary of Defense – has formally examined various options to replace the Ohio ballistic missile submarines as they retire beginning in 2027. This analysis included a variety of replacement platform options, including designs based on the highly successful Virginia-class attack submarine program and the current Ohio-class ballistic missile submarine. In the end, the Navy elected to pursue a new design that leverages the lessons from the Ohio, the Virginia advances in shipbuilding and improvements in cost-efficiency.
>
> Recently, a variety of writers have speculated that the required survivable deterrence could be achieved more cost effectively with the Virginia-based option or by restarting the Ohio-class SSBN production line. Both of these ideas make sense at face value – which is why they were included among the alternatives assessed – but the devil is in the details. When we examined the particulars, each of these options came up short in both military effectiveness and cost efficiency.
>
> **Virginia-based SSBN design with a Trident II D5 missile.** An SSBN design based on a Virginia-class attack submarine with a large-diameter missile compartment was rejected due to a wide range of shortfalls. It would:
>
> - Not meet survivability (stealth) requirements due to poor hull streamlining and lack of a drive train able to quietly propel a much larger ship
> - Not meet at-sea availability requirements due to longer refit times (since equipment is packed more tightly within the hull, it requires more time to replace, repair and retest)
> - Not meet availability requirements due to a longer mid-life overhaul (refueling needed)
> - Require a larger number of submarines to meet the same operational requirement

- Reduce the deterrent value needed to protect the country (fewer missiles, warheads at-sea)
- Be more expensive than other alternatives due to extensive redesign of Virginia systems to work with the large missile compartment (for example, a taller sail, larger control surfaces and more robust support systems)

We would be spending more money (on more ships) to deliver less deterrence (reduced at-sea warhead presence) with less survivability (platforms that are less stealthy).

Virginia-based SSBN design with a smaller missile. Some have encouraged the development of a new, smaller missile to go with a Virginia-based SSBN. This would carry forward many of the shortfalls of a Virginia-based SSBN we just discussed, and add to it a long list of new issues. Developing a new nuclear missile from scratch with an industrial base that last produced a new design more than 20 years ago would be challenging, costly and require extensive testing. We deliberately decided to extend the life of the current missile to decouple and de-risk the complex (and costly) missile development program from the new replacement submarine program. Additionally, a smaller missile means a shorter employment range requiring longer SSBN patrol transits. This would compromise survivability, require more submarines at sea and ultimately weaken our deterrence effectiveness. With significant cost, technical and schedule risks, there is little about this option that is attractive.

Ohio-based SSBN design. Some have argued that we should re-open the Ohio production line and resume building the Ohio design SSBNs. This simply cannot be done because there is no Ohio production line. It has long since been re-tooled and modernized to build state-of-the-art Virginia-class SSNs using computerized designs and modular, automated construction techniques. Is it desirable to redesign the Ohio so that a ship with its legacy performance could be built using the new production facilities? No, since an Ohio-based SSBN would:

- Not provide the required quieting due to Ohio design constraints and use of a propeller instead of a propulsor (which is the standard for virtually all new submarines)
- Require 14 instead of 12 SSBNs by reverting to Ohio class operational availability standards (incidentally creating other issues with the New START treaty limits)
- Suffer from reduced reliability and costs associated with the obsolescence of legacy Ohio system components

Once again, the end result would necessitate procuring more submarines (14) to provide the required at-sea presence and each of them would be less stealthy and less survivable against foreseeable 21st century threats.

The Right Answer: A new design SSBN that improves on Ohio: What has emerged from the Navy's exhaustive analysis is an Ohio replacement submarine that starts with the foundation of the proven performance of the Ohio SSBN, its Trident II D5 strategic weapons system and its operating cycle. To this it adds:

- Enhanced stealth as necessary to pace emerging threats expected over its service life
- Systems commonality with Virginia (pumps, valves, sonars, etc.) wherever possible, enabling cost savings in design, procurement, maintenance and logistics
- Modular construction and use of COTS equipment consistent with those used in today's submarines to reduce the cost of fabrication, maintenance and modernization. Total ownership cost reduction (for example, investing in a life-of-the-ship reactor core enables providing the same at-sea presence with fewer platforms). Although the Ohio replacement is a "new design," it is in effect an SSBN that takes the best lessons from 50 years of undersea deterrence, from the Ohio, from the Virginia, from advances in shipbuilding efficiency and maintenance, and from the stern realities of needing to provide survivable nuclear deterrence. The result is a low-risk, cost-effective platform capable of smoothly transitioning from the Ohio and delivering effective 21st century undersea strategic deterrence.[88]

APPENDIX B. EARLIER OVERSIGHT ISSUE: A DESIGN WITH 16 VS. 20 SLBM TUBES

Overview

An earlier oversight issue for Congress concerned the plan to design the SSBN(X) with 16 SLBM tubes rather than 20—one of several decisions made to reduce the estimated average procurement cost of boats 2 through 12 in the program to $5.6 billion in FY2010 dollars.[89] Some observers were concerned that designing the SSBN(X) with 16 tubes rather than 20 would create a risk that U.S. strategic nuclear forces might not have enough capability in the 2030s and beyond to fully perform their deterrent role. These observers noted that to comply with the New Start Treaty limiting strategic nuclear weapons,

DOD plans to operate in coming years a force of 14 Trident SSBNs, each with 20 operable SLBM tubes (4 of the 24 tubes on each boat are to be rendered inoperable), for a total of 240 tubes, whereas the Navy in the Ohio replacement program is planning a force of 12 SSBNs each with 16 tubes, for a total of 192 tubes, or 20% less than 240. These observers also cited the uncertainties associated with projecting needs for strategic deterrent forces out to the year 2080, when the final SSBN(X) is scheduled to leave service. These observers asked whether the plan to design the SSBN(X) with 16 tubes rather than 20 was fully supported within all parts of DOD, including U.S. Strategic Command (STRATCOM).

In response, Navy and other DOD officials stated that the decision to design the SSBN(X) with 16 tubes rather than 20 was carefully considered within DOD, and that they believe a boat with 16 tubes will give U.S. strategic nuclear forces enough capability to fully perform their deterrent role in the 2030s and beyond.

Testimony in 2011

At a March 1, 2011, hearing before the House Armed Services Committee, Admiral Gary Roughead, then-Chief of Naval Operations, stated:

> I'm very comfortable with where we're going with SSBN-X. The decision and the recommendation that I made with regard to the number of tubes—launch tubes are consistent with the new START treaty. They're consistent with the missions that I see that ship having to perform. And even though it may be characterized as a cost cutting measure, I believe it sizes the ship for the missions it will perform.[90]

At a March 2, 2011, hearing before the Strategic Forces subcommittee of the House Armed Services Committee, the following exchange occurred:

> REPRESENTATIVE TURNER:
>
> General Kehler, thank you so much for your continued thoughts and of course your leadership. One item that we had a discussion on was the triad, of looking to—of the Navy and the tube reductions of 20 to 16, as contained in other hearings on the Hill today. I would like your thoughts on the reduction of the tubes and what you see driving that, how you see it affecting our strategic posture and any other thoughts you have on that?
>
> AIR FORCE GENERAL C. ROBERT KEHLER, COMMANDER, U.S. STRATEGIC COMMAND

Thank you, Mr. Chairman. Well, first of all, sir, let me say that the—in my mind anyway, the discussion of Trident and Ohio-class replacement is really a discussion in the context of the need to modernize the entire triad. And so, first of all, I think that it's important for us to recognize that that is one piece, an important piece, but a piece of the decision process that we need to go through.

Second, the issue of the number of tubes is not a simple black-and-white answer. So let me just comment here for a minute.

First of all, the issue in my mind is the overall number of tubes we wind up with at the end, not so much as the number of tubes per submarine.

Second, the issue is, of course, we have flexibility and options with how many warheads per missile per tube, so that's another consideration that enters into this mixture.

Another consideration that is important to me is the overall number of boats and the operational flexibility that we have with the overall number of boats, given that some number will need to be in maintenance, some number will need to be in training, et cetera.

And so those and many other factors—to include a little bit of foresight here, in looking ahead to 20 years from now in antisubmarine warfare environment that the Navy will have to operate in, all of those bear on the ultimate sideways shape configuration of a follow-on to the Ohio.

At this point, Mr. Chairman, I am not overly troubled by going to 16 tubes. As I look at this, given that we have that kind of flexibility that I just laid out; given that this is an element of the triad and given that we have some decision space here as we go forward to decide on the ultimate number of submarines, nothing troubles me operationally here to the extent that I would oppose a submarine with 16 tubes.

I understand the reasons for wanting to have 20. I understand the arguments that were made ahead of me. But as I sit here today, given the totality of the discussion, I am—as I said, I am not overly troubled by 16. Now, I don't know that the gavel has been pounded on the other side of the river yet with a final decision, but at this point, I am not overly troubled by 16.[91]

At an April 5, 2011, hearing before the Strategic Forces subcommittee of the House Armed Services Committee, the following exchange occurred:

REPRESENTATIVE LARSEN:

General Benedict, we have had this discussion, not you and I, I am sorry. But the subcommittee has had a discussion in the past with regards to the Ohio-class replacement program.

The new START, though, when it was negotiated, assumed a reduction from 24 missile tubes per hole to, I think, a maximum a maximum of 20.

The current configuration [for the SSBN(X)], as I understand it, would move from 24 to 16.

Can you discuss, for the subcommittee here, the Navy's rationale for that? For moving from 24 to 16 as opposed to the max of 20?

NAVY REAR ADMIRAL TERRY BENEDICT, DIRECTOR, STRATEGIC SYSTEMS PROGRAMS (SSP):

Sir, as part—excuse me, as part of the work-up for the milestone A [review for the SSBN(X) program] with Dr. Carter in OSD, SSP supported the extensive analysis at both the OSD level as well as STRATCOM's analysis.

Throughout that process, we provided, from the SWS [strategic weapon system] capability, our perspective. Ultimately that was rolled up into both STRATCOM and OSD and senior Navy leadership and in previous testimony, the secretary of the Navy, the CNO, and General Chilton have all expressed their confidence that the mission of the future, given their perspectives, is they see the environment today can be met with 16.

And so, as the acquisition and the SWS provider, we are prepared to support that decision by leadership, sir.

REPRESENTATIVE LARSEN: Yes.

And your analysis supports—did your analysis that fed into this, did you look at specific numbers then?

REARD ADMIRAL BENEDICT:

Sir, we looked at the ability of the system, again, SSP does not look at specific targets with...

REPRESENTATIVE LARSEN:

Right. Yes, yes, yes.

REAR ADMIRAL BENEDICT:

Our input was the capability of the missile, the number of re-entry bodies and the throw weight that we can provide against those targets and based on that analysis, the leadership decision was 16, sir.[92]

At an April 6, 2011, hearing before the Strategic Forces subcommittee of the Senate Armed Services Committee, the following exchange occurred:

SENATOR SESSIONS:

Admiral Benedict, according to recent press reports, the Navy rejected the recommendations of Strategic Command to design the next generation of ballistic missile submarines with 20 missile tubes instead of opting for only 16 per boat.

What is the basis for the Navy's decision of 16? And I'm sure cost is a factor. In what ways will that decision impact the overall nuclear force structure associated with the command?

NAVY REAR ADMIRAL TERRY BENEDICT, DIRECTOR, STRATEGIC SYSTEMS PROGRAMS (SSP):

Yes, sir. SSP supported the Navy analysis, STRATCOM's analysis, as well as the OSD analysis, as we proceeded forward and towards the Milestone A decision [on the SSBN(X) program] that Dr. Carter conducted.

Based on our input, which was the technical input as the—as the director of SSP, other factors were considered, as you stated. Cost was one of them. But as the secretary, as the CNO, and I think as General Kehler submitted in their testimony, that given the threats that we see today, given the mission that we see today, given the upload capability of the D-5, and given the environment as they saw today, all three of those leaders were comfortable with the decision to proceed forward with 16 tubes, sir.

SENATOR SESSIONS:

And is that represent your judgment? To what extent were you involved—were you involved in that?

REAR ADMIRAL BENEDICT:

Sir, we were involved from technical aspects in terms of the capability of the missile itself, what we can throw, our range, our capability. And based on what we understand the capability of the D-5 today, which will be the baseline missile for the Ohio Replacement Program, as the director of SSP I'm comfortable with that decision.[93]

Section 242 Report

Section 242 of the FY2012 National Defense Authorization Act (H.R. 1540/P.L. 112-81 of December 31, 2011) required DOD to submit a report on the Ohio replacement program that includes, among other things, an assessment of various combinations of boat quantities and numbers of SLBM launch tubes per boat. The text of the section is as follows:

SEC. 242. REPORT AND COST ASSESSMENT OF OPTIONS FOR OHIO-CLASS REPLACEMENT BALLISTIC MISSILE SUBMARINE.

(a) Report Required- Not later than 180 days after the date of the enactment of this Act, the Secretary of the Navy and the Commander of the United States Strategic Command shall jointly submit to the congressional defense committees a report on each of the options described in subsection (b) to replace the Ohio-class ballistic submarine program. The report shall include the following:

(1) An assessment of the procurement cost and total life-cycle costs associated with each option.

(2) An assessment of the ability for each option to meet—

(A) the at-sea requirements of the Commander that are in place as of the date of the enactment of this Act; and

(B) any expected changes in such requirements.

(3) An assessment of the ability for each option to meet—

(A) the nuclear employment and planning guidance in place as of the date of the enactment of this Act; and

(B) any expected changes in such guidance.

(4) A description of the postulated threat and strategic environment used to inform the selection of a final option and how each option provides flexibility for responding to changes in the threat and strategic environment.

(b) Options Considered- The options described in this subsection to replace the Ohio-class ballistic submarine program are as follows:

(1) A fleet of 12 submarines with 16 missile tubes each.

(2) A fleet of 10 submarines with 20 missile tubes each.

(3) A fleet of 10 submarines with 16 missile tubes each.

(4) A fleet of eight submarines with 20 missile tubes each.

(5) Any other options the Secretary and the Commander consider appropriate.

(c) Form- The report required under subsection (a) shall be submitted in unclassified form, but may include a classified annex.

Subsection (c) above states the report "shall be submitted in unclassified form, but may include a classified annex."

The report as submitted was primarily the classified annex, with a one-page unclassified summary, the text of which is as follows (underlining as in the original):

The National Defense Authorization Act (NDAA) for Fiscal Year 2012 (FY12) directed the Secretary of the Navy and the Commander of U.S. Strategic Command (USSTRATCOM) to jointly submit a report to the congressional defense committees comparing four different options for the OHIO Replacement (OR) fleet ballistic missile submarine (SSBN) program. Our assessment considered the current operational requirements and guidance. The four SSBN options analyzed were:

1. 12 SSBNs with 16 missile tubes each
2. 10 SSBNs with 20 missile tubes each
3. 10 SSBNs with 16 missile tubes each
4. 8 SSBNs with 20 missile tubes each

The SSBN force continues to be an integral part of our nuclear Triad and contributes to deterrence through an assured second strike capability that is survivable, reliable, and credible. The number of SSBNs and their combined missile tube capacity are important factors in our flexibility to respond to changes in the threat and uncertainty in the strategic environment.

We assessed each option against the ability to meet nuclear employment and planning guidance, ability to satisfy at-sea requirements, flexibility to respond to future changes in the postulated threat and strategic environment, and cost. In general, options with more SSBNs can be adjusted downward in response to a diminished threat; however, options with less SSBNs are more difficult to adjust upward in response to a growing threat.

Clearly, a smaller SSBN force would be less expensive than a larger force, but for the reduced force options we assessed, they fail to meet current at-sea and nuclear employment requirements, increase risk in force survivability, and limit flexibility in response to an uncertain strategic future. Our assessment is the program of record, 12 SSBNs with 16 missile tubes each, provides the best balance of performance, flexibility, and cost meeting commander's requirements while supporting the Nation's strategic deterrence mission goals and objectives.

The classified annex contains detailed analysis that is not releasable to the public.[94]

End Notes

[1] CRS Report RL33640, U.S. Strategic Nuclear Forces: Background, Developments, and Issues, by Amy F. Woolf.

[2] In the designations SSN, SSGN, SSBN, and SSBN(X), the SS stands for submarine, N stands for nuclear-powered (meaning the ship is powered by a nuclear reactor), G stands for guided missile (such as a cruise missile), B stands for ballistic missile, and (X) means the design of the ship has not yet been determined.

As shown by the "Ns" in SSN, SSGN, and SSBN, all U.S. Navy submarines are nuclear-powered. Other navies operate non-nuclear powered submarines, which are powered by energy sources such as diesel engines. A submarine's use of nuclear or non-nuclear power as its energy source is not an indication of whether it is armed with nuclear weapons—a nuclear-powered submarine can lack nuclear weapons, and a non-nuclear-powered submarine can be armed with nuclear weapons.

[3] These missions include covert intelligence, surveillance, and reconnaissance (ISR), much of it done for national-level (as opposed to purely Navy) purposes; covert insertion and recovery of special operations forces (SOF); covert strikes against land targets with the Tomahawk cruise missiles; covert offensive and defensive mine warfare; anti-submarine warfare (ASW); and anti-surface ship warfare. The Navy's four SSGNs, which are converted former SSBNs, can carry larger numbers of Tomahawks and SOF personnel than can the SSNs. SSGN operations consequently may focus more strongly on Tomahawk and SOF missions than do SSN operations. For more on the Navy's SSNs and SSGNs, see CRS Report RL32418, Navy Virginia (SSN-774) Class Attack Submarine Procurement: Background and Issues for Congress, by Ronald O'Rourke, and CRS Report RS21007, Navy Trident

Submarine Conversion (SSGN) Program: Background and Issues for Congress, by Ronald O'Rourke.

[4] The Navy's non-strategic nuclear weapons—meaning all of the service's nuclear weapons other than submarine-launched ballistic missiles (SLBMs)—were removed from Navy surface ships and submarines under a unilateral U.S. nuclear initiative announced by President George H. W. Bush in September 1991. The initiative reserved a right to rearm SSNs at some point in the future with nuclear-armed Tomahawk land attack missiles (TLAM-Ns) should conditions warrant. Navy TLAM-Ns were placed in storage to support this option. DOD's report on the 2010 Nuclear Posture Review (NPR), released on April 6, 2010, states that the United States will retire the TLAM-Ns. (Department of Defense, Nuclear Posture Review Report, April 2010, pp. xiii and 28.)

[5] SSBNs, like other Navy submarines, are also equipped with horizontal torpedo tubes in the bow for firing torpedoes or other torpedo-sized weapons.

[6] This informal name is a reference to the large boom that would be made by the detonation of an SLBM nuclear warhead.

[7] U.S. Navy, Report to Congress on Annual Long-Range Plan for Construction of Naval Vessels for FY 2011, February 2010, p. 15.

[8] Department of Defense, Nuclear Posture Review Report, April 2010, p. 22. The next sentence in the report states: "Today, there appears to be no viable near or mid-term threats to the survivability of U.S. SSBNs, but such threats—or other technical problems—cannot be ruled out over the long term." The report similarly states on page 23: "Today, there appears to be no credible near or mid-term threats to the survivability of U.S. SSBNs. However, given the stakes involved, the Department of Defense will continue a robust SSBN Security Program that aims to anticipate potential threats and develop appropriate countermeasures to protect current and future SSBNs."

[9] For more on the SSGN conversion program, see CRS Report RS21007, Navy Trident Submarine Conversion (SSGN) Program: Background and Issues for Congress, by Ronald O'Rourke.

[10] Statement of Rear Admiral Stephen Johnson, USN, Director, Strategic Systems Programs, Before the Subcommittee on Strategic Forces of the Senate Armed Services Committee [on] FY2011 Strategic Systems, March 17, 2010, p. 4.

[11] The larger size of the Ohio-class design also reflects a growth in size over time in U.S. submarine designs due to other reasons, such as providing increased interior volume for measures to quiet the submarine acoustically, so as to make it harder to detect.

[12] Although the SLBMs on UK SSBNs are U.S.-made, the nuclear warheads on the missiles are of UK design and manufacture.

[13] A March 18, 2010, report by the UK Parliament's House of Commons Foreign Affairs Committee stated:

During the Cold War, the UK's nuclear co-operation with the United States was considered to be at the heart of the [UK-U.S.] 'special relationship'. This included the 1958 Mutual Defence Agreement, the 1963 Polaris Sales Agreement (PSA) (subsequently amended for Trident), and the UK's use of the US nuclear test site in Nevada from 1962 to 1992. The co-operation also encompassed agreements for the United States to use bases in Britain, with the right to store nuclear weapons, and agreements for two bases in Yorkshire (Fylingdales and Menwith Hill) to be upgraded to support US missile defence plans.

In 1958, the UK and US signed the Mutual Defence Agreement (MDA). Although some of the appendices, amendments and Memoranda of Understanding remain classified, it is known that the agreement provides for extensive co-operation on nuclear warhead and reactor technologies, in particular the exchange of classified information concerning nuclear weapons to improve design, development and fabrication capability. The agreement also provides for the transfer of nuclear warhead-related materials. The agreement was renewed in 2004 for another ten years.

The other major UK-US agreement in this field is the 1963 Polaris Sales Agreement (PSA) which allows the UK to acquire, support and operate the US Trident missile system. Originally signed to allow the UK to acquire the Polaris Submarine Launched Ballistic Missile (SLBM) system in the 1960s, it was amended in 1980 to facilitate purchase of the Trident I (C4) missile and again in 1982 to authorise purchase of the more advanced Trident II (D5) in place of the C4. In return, the UK agreed to formally assign its nuclear forces to the defence of NATO, except in an extreme national emergency, under the terms of the 1962 Nassau Agreement reached between President John F. Kennedy and Prime Minister Harold Macmillan to facilitate negotiation of the PSA.

Current nuclear co-operation takes the form of leasing arrangements of around 60 Trident II D5 missiles from the US for the UK's independent deterrent, and long-standing collaboration on the design of the W76 nuclear warhead carried on UK missiles. In 2006 it was revealed that the US and the UK had been working jointly on a new 'Reliable Replacement Warhead' (RRW) that would modernise existing W76-style designs. In 2009 it emerged that simulation testing at Aldermaston on dual axis hydrodynamics experiments had provided the US with scientific data it did not otherwise possess on this RRW programme.

The level of co-operation between the two countries on highly sensitive military technology is, according to the written submission from Ian Kearns, "well above the norm, even for a close alliance relationship". He quoted Admiral William Crowe, the former US Ambassador to London, who likened the UK-US nuclear relationship to that of an iceberg, "with a small tip of it sticking out, but beneath the water there is quite a bit of everyday business that goes on between our two governments in a fashion that's unprecedented in the world." Dr Kearns also commented that the personal bonds between the US/UK scientific and technical establishments were deeply rooted.

(House of Commons, Foreign Affairs Committee, Sixth Report Global Security: UK-US Relations, March 18, 2010, paragraphs 131-135; http://www.publications.parliament.uk /pa/cm200910/ cmselect/cmfaff/114/11402.htm; paragraphs 131-135 are included in the section of the report available at http://www.publications.parliament.uk /pa/cm200910/cmselect/cmfaff/114/11406.htm.)

See also "U.K. Stays Silent on Nuclear-Arms Pact Extension with United States," Global Security Newswire (www.nti.org/gsn), July 30, 2014.

14 Statement of Rear Admiral Stephen Johnson, USN, Director, Strategic Systems Programs, Before the Subcommittee on Strategic Forces of the Senate Armed Services Committee [on] FY2011 Strategic Systems, March 17, 2010, p. 6.

15 Securing Britain in an Age of Uncertainty: The Strategic Defence and Security Review, Presented to Parliament by the Prime Minister by Command of Her Majesty, October 2010, p. 39.

16 Securing Britain in an Age of Uncertainty: The Strategic Defence and Security Review, Presented to Parliament by the Prime Minister by Command of Her Majesty, October 2010, pp. 5, 38-39. For more on the UK's Successor SSBN program as it existed prior to the October 2010 UK defense and security review report, see Richard Scott, "Deterrence At A Discount?" Jane's Defence Weekly, December 23, 2009: 26-31.

[17] Securing Britain in an Age of Uncertainty: The Strategic Defence and Security Review, Presented to Parliament by the Prime Minister by Command of Her Majesty, October 2010, p. 39.

[18] PWR3 means pressurized water reactor, design number 3. U.S. and UK nuclear-powered submarines employ pressurized water reactors. Earlier UK nuclear powered submarines are powered by reactor designs that the UK designated PWR-2 and PWR-1. For an article discussing the PWR3 plant, see Richard Scott, "Critical Mass: Re-Energising the UK's Naval Nuclear Programme," Jane's International Defence Review, July 2014: 42-45, 47.

[19] Sam LaGrone and Richard Scott, "Strategic Assets: Deterrent Plans Confront Cost Challenges," Jane's Navy International, December 2011: 17 and 18.

[20] Source: Email to CRS from Navy Office of Legislative Affairs, June 25, 2012. See also Jon Rosamond, "Next Generation U.K. Boomers Benefit from U.S. Relationship," USNI News (http://news.usni.org), December 17, 2014.

[21] Sam LaGrone and Richard Scott, "Strategic Assets: Deterrent Plans Confront Cost Challenges," Jane's Navy International, December 2011: 19.

[22] In February 2007, the commander of U.S. Strategic Command (STRATCOM) commissioned a task force to support an anticipated Underwater Launched Missile Study (ULMS). On June 8, 2007, the Secretary of the Navy initiated the ULMS. Six days later, the commander of STRATCOM directed that a Sea Based Strategic Deterrent (SBSD) capability-based assessment (CBA) be performed. In July 2007, the task force established by the commander of STRATCOM provided its recommendations regarding capabilities and characteristics for a new SBSD. (Source: Navy list of key events relating to the ULMS and SBSD provided to CRS and the Congressional Budget Office (CBO) on July 7, 2008.)

[23] On February 14, 2008, the SBSD ICD was approved for joint staffing by the Navy's Resources and Requirements Review Board (R3B). On April 29, 2008, the SBSD was approved by DOD's Functional Capabilities Board (FCB) to proceed to DOD's Joint Capabilities Board (JCB). (Source: Navy list of key events relating to the ULMS and SBSD provided to CRS and CBO on July 7, 2008.)

[24] Navy briefing to CRS and CBO on the SBSD program, July 6, 2009.

[25] Navy briefing to CRS and CBO on the SBSD program, July 6, 2009.

[26] An August 2008 press report states that the program office, called PMS-397, "was established within the last two months." (Dan Taylor, "Navy Stands Up Program Office To Manage Next-Generation SSBN," Inside the Navy, August 17, 2008.

[27] "Going Ballistic," Defense Daily, September 22, 2008, p. 1.

[28] Department of Defense Fiscal Year (FY) 2012 Budget Estimates, Navy, Justification Book Volume 2, Research, Development, Test & Evaluation, Navy Budget Activity 4, entry for PE0603561N, Project 3220 (pdf page 345 of 888).

[29] Source: Email from Navy Office of Legislative Affairs to CRS, February 3, 2011.

[30] Navy budget submissions show that Ohio-class mid-life nuclear refueling overhauls have contract-award-to-delivery periods generally ranging from 47 months to 50 months.

[31] Source: Navy update briefing on Ohio replacement program to CRS and CBO, September 17, 2012. See also "Navy Responds to Debate Over the Size of the SSBN Force," Navy Live, May 16, 2013, accessed July 26, 2013, at http://navylive.dodlive.mil/2013/05/16/navy-responds-to-debate-over-the-size-of-the-ssbn-force/ and Richard Breckenridge, "SSBN Force Level Requirements: It's Simply a Matter of Geography," Navy Live, July 19, 2013, accessed July 26, 2013, at http://navylive.dodlive.mil/2013/07/19/ssbn-force-level-requirements-

[32] Source: Navy update briefing on Ohio replacement program to CRS and CBO, September 17, 2012. A September 28, 2012, press report similarly quotes Rear Admiral Barry Bruner, the Navy's director of undersea warfare, as stating that "During this time frame, no major SSBN overhauls are planned, and a force of 10 SSBNs will support current at-sea presence requirements," and that "This provides a low margin to compensate for unforeseen issues that may result in reduced SSBN availability. The reduced SSBN availability during this time frame reinforces the importance of remaining on schedule with the Ohio Replacement program to meet future strategic requirements. As the Ohio Replacement ships begin their mid-life overhauls in 2049, 12 SSBNs will be required to offset ships conducting planned maintenance." (Michael Fabey, U.S. Navy Defends Boomer Submarine Replacement Plans," Aerospace Daily & Defense Report, September 28, 2012: 3.)

[33] For more on the Virginia-class program, see CRS Report RL32418, Navy Virginia (SSN-774) Class Attack Submarine Procurement: Background and Issues for Congress, by Ronald O'Rourke.

[34] Rear Admiral David Johnson, briefing to Naval Submarine League Annual Symposium [on] Expanding Undersea Dominance, October 23, 2014, briefing slide 19.

[35] As mentioned earlier (see "Current Ohio-Class SSBNs"), the Ohio-class boats receive a mid-life nuclear refueling overhaul, called an Engineered Refueling Overhaul (ERO), which includes both a nuclear refueling and overhaul work on the ship that is not related to the nuclear refueling.

[36] U.S. Navy, Report to Congress on Annual Long-Range Plan for Construction of Naval Vessels for FY 2011, February 2010, p. 5. The two most recent classes of SSNs—the Seawolf (SSN-21) and Virginia (SSN-774) class boats—are built with cores that are expected to be sufficient for their entire 33-year expected service lives.

[37] Source: Rear Admiral David Johnson, briefing to Naval Submarine League Annual Symposium [on] Expanding Undersea Dominance, October 23, 2014, briefing slide 19. See also the spoken testimony of Admiral Kirkland Donald, Deputy Administrator for Naval Reactors, and Director, Naval Nuclear Propulsion, National Nuclear Security Administration, at a March 30, 2011, hearing before the Strategic Forces Subcommittee of the Senate Armed Services Committee, as shown in the transcript of the hearing, and Dave Bishop, "What Will Follow the Ohio Class?" U.S. Naval Institute Proceedings, June 2012: 31; and Sam LaGrone and Richard Scott, "Strategic Assets: Deterrent Plans Confront Cost Challenges," Jane's Navy International, December 2011: 16.

[38] Beam is the maximum width of a ship. For Navy submarines, which have cylindrical hulls, beam is the diameter of the hull.

[39] Dave Bishop, "What Will Follow the Ohio Class?" U.S. Naval Institute Proceedings, June 2012: 31. (Bishop was program manager for the Ohio replacement program.) See also Sam LaGrone and Richard Scott, "Strategic Assets: Deterrent Plans Confront Cost Challenges," Jane's Navy International, December 2011: 15 and 16.

[40] Sydney J. Freedberg, "Navy Seeks Sub Replacement Savings: From NASA Rocket Boosters To Reused Access Doors," Breaking Defense (http://breakingdefense.com), April 7, 2014.

[41] Navy information paper on Ohio replacement program dated August 11, 2014, provided to CBO and CRS on August 11, 2014.

[42] U.S. Navy, Report to Congress on Annual Long-Range Plan for Construction of Naval Vessels for FY 2011, February 2010, p. 24. See also Mike McCarthy, "Navy Striving To Reduce Detectability Of Next Boomers," Defense Daily, February 6, 2015: 1.

[43] The term X-stern means that the steering and diving fins at the stern of the ship are, when viewed from the rear, in the diagonal pattern of the letter X, rather than the vertical-and horizontal pattern of a plus sign (which is referred to as a cruciform stern).

[44] The common submarine radio room is a standardized (i.e., common) suite of submarine radio room equipment that is being installed on other U.S. Navy submarines.

[45] Dave Bishop, "What Will Follow the Ohio Class?" U.S. Naval Institute Proceedings, June 2012: 31. See also Sam LaGrone and Richard Scott, "Strategic Assets: Deterrent Plans Confront Cost Challenges," Jane's Navy International, December 2011: 15 and 16. The X-stern is also shown in Rear Admiral David Johnson, briefing to Naval Submarine League Annual Symposium [on] Expanding Undersea Dominance, October 23, 2014, briefing slide 19.

[46] Government Accountability Office, Defense Acquisitions[:] Assessments of Selected Weapons Programs, GAO-15- 342SP, March 2015, p. 148.

[47] Source: Navy information paper dated February 3, 2015, provided to CRS and CBO on February 24, 2015. See also Statement of the Honorable Sean J. Stackley, Assistant Secretary of the Navy (Research, Development and Acquisition) and Vice Admiral Joseph P. Mulloy, Deputy Chief of Naval Operations for Integration of Capabilities and Resources and Lieutenant General Kenneth J. Glueck, Jr., Deputy Commandant, Combat Development and Integration & Commanding General, Marine Corps Combat Development Command, Before the Subcommittee on Seapower and Projection Forces of the House Armed Services Committee on Department of the Navy Seapower and Projection Forces Capabilities, February 25, 2015, p. 7.

[48] U.S. Navy, Report to Congress on Annual Long-Range Plan for Construction of Naval Vessels for FY 2011, February 2010, p. 20.

[49] Christopher J. Castelli, "DOD: New Nuclear Subs Will Cost $347 Billion To Acquire, Operate," Inside the Navy, February 21, 2011; Elaine M. Grossman, "Future U.S. Nuclear-Armed Vessel to Use Attack-Submarine Technology," Global Security Newswire, February 24, 2011; Jason Sherman, "Navy Working To Cut $7.7 Billion From Ohio Replacement Program," Inside the Navy, February 28, 2011. See also Christopher J. Castelli, "DOD Puts 'Should-Cost' Pressure On Major Weapons Programs," Inside the Navy, May 2, 2011.

[50] Lee Hudson, "Navy Continues Working Toward SSBN(X) Cost Target, Slashes $360M," Inside the Navy, January 26, 2015.

[51] Dave Bishop, "Two Years In And Ground Strong, The Ohio Replacement Program," Undersea Warfare, Spring 2012: 5; Megan Eckstein, "Ohio-Replacement Sub Technology To Drop O&S Costs To $110M A Year," Inside the Navy, March 1, 2013.

[52] Statement of Rear Admiral Stephen Johnson, USN, Director, Strategic Systems Programs, Before the Subcommittee on Strategic Forces of the Senate Armed Services Committee [on] FY2011 Strategic Systems, March 17, 2010, p. 6, which states: "The OHIO Replacement programs includes the development of a common missile compartment that will support both the OHIO Class Replacement and the successor to the UK Vanguard Class."

[53] Government Accountability Office, Defense Acquisitions[:] Assessments of Selected Weapon Programs, GAO-10- 388SP, March 2010, p. 152.

[54] Government Accountability Office, Defense Acquisitions[:] Assessments of Selected Weapon Programs, GAO-11- 233SP, March 2011, p. 147.

[55] Sam LaGrone and Richard Scott, "Deterrent Decisions: US and UK Wait on Next Steps for SSBN Replacements," Jane's Navy International, May 2010, pp. 10-11.

[56] Lee Hudson, "Navy Continues Working Toward SSBN(X) Cost Target, Slashes $360M," Inside the Navy, January 26, 2015.

[57] For more on the CVN-78 program, see CRS Report RS20643, Navy Ford (CVN-78) Class Aircraft Carrier Program: Background and Issues for Congress, by Ronald O'Rourke.

[58] For more on the LPD-17 program, see CRS Report RL34476, Navy LPD-17 Amphibious Ship Procurement: Background, Issues, and Options for Congress, by Ronald O'Rourke.

[59] For more on the LCS program, see CRS Report RL33741, Navy Littoral Combat Ship (LCS)/Frigate Program: Background and Issues for Congress, by Ronald O'Rourke.

[60] Congressional Budget Office, An Analysis of the Navy's Fiscal Year 2015 Shipbuilding Plan, December 2014, pp. 23-25.

[61] Statement of Admiral Jonathan Greenert, U.S. Navy, Chief of Naval Operations, Before the House Subcommittee on Defense, Committee on Appropriations, on FY 2016 Department of the Navy Posture, February 26, 2015, p. 5.

[62] Statement of the Honorable Sean J. Stackley, Assistant Secretary of the Navy (Research, Development and Acquisition) and Vice Admiral Joseph P. Mulloy, Deputy Chief of Naval Operations for Integration of Capabilities and Resources and Lieutenant General Kenneth J. Glueck, Jr., Deputy Commandant, Combat Development and Integration & Commanding General, Marine Corps Combat Development Command, Before the Subcommittee on Seapower and Projection Forces of the House Armed Services Committee on Department of the Navy Seapower and Projection Forces Capabilities, February 25, 2015, p. 8.

[63] Statement of Admiral Jonathan Greenert, U.S. Navy, Chief of Naval Operations, Before the House Armed Services Committee on Planning for Sequestration in FY 2014 and Perspectives of the Military Services on the Strategic Choices and Management Review, September 18, 2013, p. 10.

[64] Transcript of hearing. (Spoken remarks of Rear Admiral Richard Breckenridge. The other witness at the hearing was Rear Admiral David Johnson).

[65] Transcript of hearing. (Spoken remarks of Rear Admiral Richard Breckenridge.)

[66] Transcript of hearing. (Spoken remarks of Rear Admiral Richard Breckenridge.)

[67] Transcript of hearing (Spoken remarks of Rear Admiral Richard Breckenridge.) Regarding supplemental funding for the 41 earlier SSBNs, Breckenridge stated:

The—just a little backstep and history to talk about the two other times that we've had to, as a nation, build the strategic deterrent. So in—in the '60s we built 41 SSBNs; they were called the 41 For Freedom. We did that in a seven-year period, which again is just an incredible—only in America could you go ahead and put out 41 ballistic missile submarines in a seven-year period.

There was an impact to other shipbuilding accounts at that time, but the priority was such for national survival that we had to go ahead and—and make that a—an imperative and a priority.

There was a supplement to the Navy's top line at that time when we—when we fielded the class, but it did leave—cast quite a shadow over the rest of the shipbuilding in the '60s.

We recapitalized those 41 For Freedom with 18 Ohio-class SSBNs in the '80s. It was the Reagan years. There was a major naval buildup. And underneath the umbrella of that buildup we were able to afford as a nation the recapitalization of building 18 SSBNs.

See also Joseph Tofalo, "The Value of Sea Based Strategic Deterrence," Navy Live (http://navylive.dodlive.mil), July 31, 2014.

[68] Transcript of hearing. (Spoken remarks of Rear Admiral Richard Breckenridge.) See also Christopher J. Castelli, "Admiral: DOD Likely To Support SSBN(X) Supplemental Funding," Inside the Navy, November 11, 2013.

[69] Department of Defense, National Defense Budget Estimates For FY 2015, April 2014, Table 6-4, "Department of Defense TOA by Program," page 98. See also Table 6-5 on page 100,

which presents the same data in constant FY2015 dollars. The other 10 program areas in addition to strategic forces are general purpose forces; C3, intelligence and space; mobility forces; guard and reserve forces; research and development; central supply and management; training, medical and other; administration and associated; support of other nations; and special operations forces. (A 12th category—other—shows relatively small amounts of funding.)

[70] For more on MYP and block buy contracting, see CRS Report R41909, Multiyear Procurement (MYP) and Block Buy Contracting in Defense Acquisition: Background and Issues for Congress, by Ronald O'Rourke and Moshe Schwartz

[71] See CRS Report RL32418, Navy Virginia (SSN-774) Class Attack Submarine Procurement: Background and Issues for Congress, by Ronald O'Rourke.

[72] Government Accountability Office, Defense Acquisitions[:] Assessments of Selected Weapons Programs, GAO-14-340SP, March 2014, p. 141.

[73] Lee Hudson, "Navy SSBN(X) Acquisition Strategy Will Not Be Finalized Until Fall 2016," Inside the Navy, September 8, 2014.

[74] Under split funding, a boat's procurement cost is divided into two parts, or increments. The first increment would be provided in the fiscal year that the boat is procured, and the second would be provided the following fiscal year.

[75] Procuring one SSBN(X) every two years rather than at the Navy's planned rate of one per year could result in a loss of learning at the shipyard in moving from production of one SSBN to the next.

[76] The Navy, with congressional support, currently uses split funding to procure large-deck amphibious assault ships (i.e., LHAs). The Navy currently is permitted by Congress to use four-year incremental funding for procuring the first three Ford (CVN-78) class carriers (i.e., CVN-78, CVN-79, and CVN-80); the authority was granted in §121 of the FY2007 defense authorization act [H.R. 5122/P.L. 109-364 of October 17, 2006]).

[77] Congress funded the procurement of two aircraft carriers (CVNs 72 and 73) in FY1983, and another two (CVNs 74 and 75) in FY1988. Although CVN-73 was funded in FY1983, it was built on a schedule consistent with a carrier funded in FY1985; although CVN-75 was funded in FY1988, it was built on a schedule consistent with a carrier funded in FY1990 or FY1991.

[78] Christopher J. Castelli, "White House Opposes Incremental Funding For Multibillion-Dollar Sub," Inside the Navy, December 19, 2011.

[79] Congressional Budget Office, Options for Reducing the Deficit: 2014 to 2023, November 2013, pp. 68-69.

[80] See, for example, Congressional Budget Office, Rethinking the Trident Force, July 1993, 78 pp.; and Congressional Budget Office, Budget Options, March 2000, p. 62.

[81] Debt, Deficits, and Defense, A Way Forward[:] Report of the Sustainable Defense Task Force, June 11, 2010, pp. 19-20.

[82] Benjamin H. Friedman and Christopher Preble, Budgetary Savings from Military Restraint, Washington, Cato Institute, September 23, 2010 (Policy Analysis No. 667), p. 8.

[83] Strategic Agility: Strong National Defense for Today's Global and Fiscal Realities, Stimson, Washington, DC, 2013, p. 29. (Sponsored by the Peter G. Peterson Foundation, Prepared by Stimson, September 2013.)

[84] For further discussion, see CRS Report RL33640, U.S. Strategic Nuclear Forces: Background, Developments, and Issues, by Amy F. Woolf.

[85] Lee Hudson, "Navy SSBN(X) Acquisition Strategy Will Not Be Finalized Until Fall 2016," Inside the Navy, September 8, 2014.

[86] Under the joint-production arrangement for Virginia-class boats, GD/EB builds certain parts of each boat, NNS builds certain other parts of each boat, and the two yards take turns building the reactor compartment and performing final assembly work. GD/EB is the final assembly yard for the first Virginia-class boat, the third one, and so on, while NNS is the final assembly yard for the second boat, the fourth one, and so on. The arrangement provides a roughly 50- 50 split in profits between the two firms for the production of Virginia-class SSNs. The agreement governing the joint-production arrangement cannot be changed without the consent of both firms. Virginia-class SSNs are the first U.S. nuclear-powered submarines to be built jointly by two shipyards; all previous U.S. nuclear-powered submarines were built under the more traditional approach of building an entire boat within a single yard.

The Virginia-class joint-production arrangement was proposed by the two shipyards, approved by the Navy, and then approved by Congress as part of its action on the FY1998 defense budget. A principal goal of the arrangement is to preserve submarine-construction skills at two U.S. shipyards while minimizing the cost of using two yards to build a class of submarines that is procured at a relatively low rate of one or two boats per year. Preserving submarine-construction skills at two yards is viewed as a hedge against the possibility of operations at one of the yards being disrupted by a natural or man-made disaster.

The joint-production arrangement is more expensive than single-yard strategy of building all Virginia-class boats at one shipyard (in part because the joint-production strategy splits the learning curve for reactor compartment construction and final assembly work on Virginia-class SSNs), but it is less expensive than a separate-yard strategy of building complete Virginia-class separately at both yards (in part because a separate-construction strategy splits the learning curve for all aspects of Virginia-class construction work, and because, in the absence of other submarine-construction work, a procurement rate of one or two Virginia-class boats per year is viewed as insufficient to sustain a meaningful competition between the two yards for contracts to build the boats).

[87] The agreement governing the joint-production arrangement for the Virginia class cannot be changed without the consent of both yards.

[88] "Facts We Can Agree Upon About Design of Ohio Replacement SSBN," Navy Live, accessed July 3, 2013, at http://navylive.dodlive.mil/2013/06/26/facts-we-can-agree-upon-about-design-of-ohio-replacement-ssbn/.

[89] At a March 30, 2011, hearing before the Strategic Forces subcommittee of the Senate Armed Services Committee, Admiral Kirkland Donald, Deputy Administrator for Naval Reactors and Director, Naval Nuclear Propulsion, National Nuclear Security Administration, when asked for examples cost efficiencies that are being pursued in his programs, stated:

The—the Ohio replacement [program] has been one that we've obviously been focused on here for—for several years now. But in the name of the efficiencies, and one of the issues as we work through the Defense Department's acquisition process, we were the first program through that new process that Dr. [Aston] Carter [the DOD acquisition executive] headed up.

But we were challenged to—to drive the cost of that ship down, and as far as our part was concerned, one of the key decisions that was made that—that helped us in that regard was a decision to go from 20 missile tubes to 16 missile tubes, because what that allowed us to do was to down rate the—the propulsion power that was needed, so obviously, it's a – it's a small[er] the reactor that you would need.

But what it also allowed us to do was to go back [to the use of existing components]. The size [of the ship] fell into the envelope where we could go back and use components that we had already designed for the Virginia class [attack submarines] and bring those into this design, not have to do it over again, but several of the mechanical components, to use those over again.

And it enabled us to drive the cost of that propulsion plant down and rely on proven technology that's—pumps and valves and things like that don't change like electronics do.

So we're pretty comfortable putting that in ship that'll be around 'til 2080. But we were allowed to do that.

(Source: Transcript of hearing.)

[90] Source: Transcript of hearing.

[91] Source: Transcript of hearing.

[92] Source: Transcript of hearing.

[93] Source: Transcript of hearing.

[94] Report and Cost Assessment of Options for OHIO-Class Replacement Ballistic Missile Submarine, Unclassified Summary, received from Navy Legislative Affairs Office, August 24, 2012. See also Christopher J. Castelli, "Classified Navy Assessment On SSBN(X) Endorses Program Of Record," Inside the Navy, September 10, 2012.

INDEX

#

A

B

C

K

L

M

N

O

P

R

S

T

U

V

W